Florian Ion & Relly Victoria
PETRESCU

MECHATRONICS

-

SERIAL AND PARALLEL SYSTEMS

CREATE SPACE PUBLISHER USA 2013

Scientific reviewer:

Dr. Veturia CHIROIU
Honorific member of
Technical Sciences Academy of Romania (ASTR)
PhD supervisor in Mechanical Engineering

Copyright

Title book: Mechatronics – Serial and Parallel Systems

Authors book: Florian Ion PETRESCU & Relly Victoria PETRESCU

© 2011-2013 Florian PETRESCU

petrescuflorian@yahoo.com

ALL RIGHTS RESERVED. This book contains material protected under International and Federal Copyright Laws and Treaties. Any unauthorized reprint or use of this material is prohibited. No part of this book may be reproduced or transmitted in any form or by any means, electronic or mechanical, including photocopying, recording, or by any information storage and retrieval system without express written permission from the author / publisher.

ISBN 978-1-4942-4152-0

WELCOME

THE THREE FUNDAMENTAL

LAWS OF ROBOTICS:

1. The robot must not harm humans or through inaction allow anything to happen to a human being.

2. The robot must obey human commands, but only when they do not contradict the Law 1.

3. The robot must protect its existence, but only when self-care law does not contradict the law of 1 or 2.

Moving mechanical structures are used increasingly in almost all vital sectors of humanity. The robots are able to process integrated circuits sizes micro and nano, on which the man they can be seen even with electron microscopy. Dyeing parts in toxic environments, working in chemical and radioactive environments, or at depths and pressures at the bottom of huge oceans, or even cosmic space conquest and visiting exo-planets, are now possible, and were turned into from the dream in reality, because mechanical platforms sequential gearbox.

Robots were developed and diversified, different aspects, but to-day, they start to be directed on two major categories: systems serial and parallel systems. Parallel systems are more solid, but more difficult to designed and handled, which serial systems were those which have developed the most. In medical operations or radioactive environments are preferred mobile systems parallel to their high accuracy positioning.

CONTENT

Welcome... 003

Content..004

First Part: Serial Systems

Chapter 01_Moving Mechanical Systems Structure, Serial 005

Chapter 02_The MP-3R Inverse Kinematics... 011

Chapter 03_The Structure and Direct Kinematics..031

Chapter 04_ Inverse Kinematics of Moving (Serial) Mechanical Systems by a Geometric Method... 046

Chapter 05 Inverse Kinematics to the Anthropomorphic Robots, by the Trigonometric Method..056

Chapter 06_The Transition from 2R System to the 3R System...074

Chapter 07_Kinetostatic and Dynamics 080

Second Part: Parallel Systems

Chapter 08_ Parallel Moving Mechanical Systems 098

Chapter 01_Moving Mechanical Systems Structure, Serial

The most commonly used serial structures in the past 20-30 years are those of type 3R, 4R, 5R, 6R, as constituents essential basic kinematic chain 3R, robot antropomorf (RRR), where main rotation around a vertical axis, results in the construction.

Then there are a basic kinematic chain which has two revolutions kinematic (two actuators, i.e. two motors) who work permanently in one plane, and immediately after main support which supports and rotates vertically complete assembly.

This the basic structure, 3R, it comes out at all the robots serial connectors manufactured on the principle of rotations. Vertical Bracket it is always the same, but the drive train which follows with two turns situated in a plane can be positioned vertically (most often; the robots anthropomorphic, fig. 1b), or horizontally (case robots scale, fig. 1a).

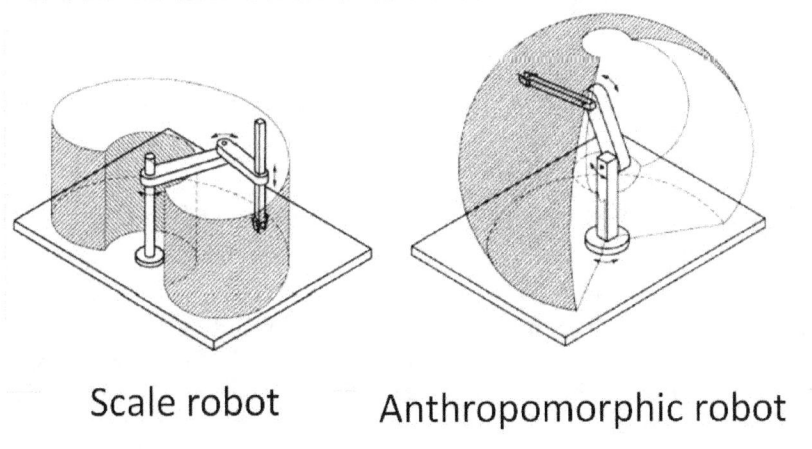

Scale robot   Anthropomorphic robot

a   b
Fig. 1. *The basic structures 3R (a-structure scale; b-structure anthropomorphic)*

It can thus pass from the study of spatial movement, which is more difficult to study planar motion, motion base for all robots and manipulators series with rotating movements.

Movement flat, vertical or horizontal, is much easier than studying space, taking advantage of easy integration into space part.

Next we illustrate the basic structure existing in several series platform rotation as the most generalized (most common) today.

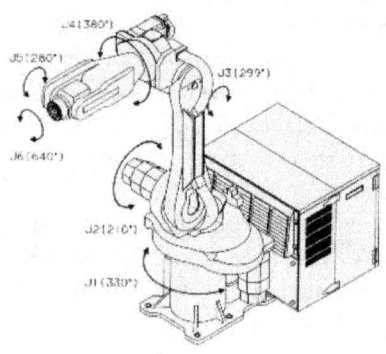

Fig. 2. *Structure 6R (anthropomorphic)*

In this basic model (3R) were further developed 6R robots today (fig. 2, based only on rotation, using only the electric motors that drive light, compact), they have a greater rigidity and flexibility while maintaining penetration patterns 3R, 4R and 5R.

Almost all major companies today build models 6R (which they improve continuously).

Why they have imposed today these models of robots (after decades of diversity was the word of order?); may and because of the need for standardization, or to find a common solution, after a huge portion (however are not yet the only ones robots use of category serialilor, but they also have the widest spread).

The six turns (full elimination of translatiilor, who bring many disadvantages due to coupler T itself) are operating easier, faster, with higher yield, more reliable, more compact and more secure; rotations, remain all first three, the other three turns (additional) having the role of positioning may well end device, the endeffector. Result and here that the baseline study (required) it is still not for a 3R.

This can also be seen in the latest models of the various firms producing robots (fig. 3, Kawasaki, Romat, Fanuc, Motoman, Kuka, etc). And structures used inside cells finding sequential gearbox are constructed generally in a similar way.

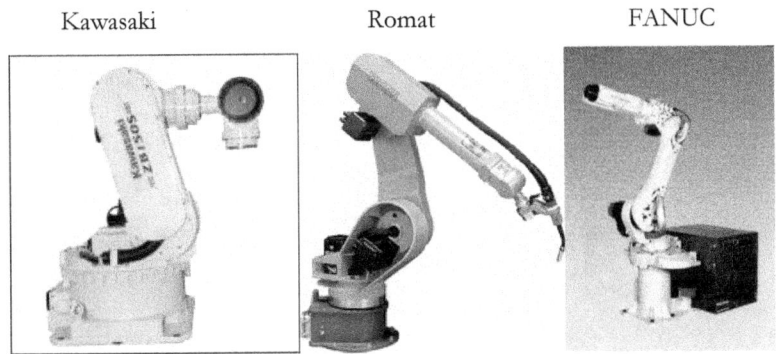
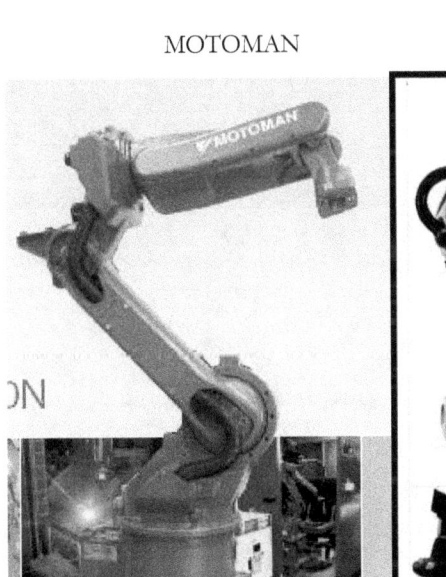

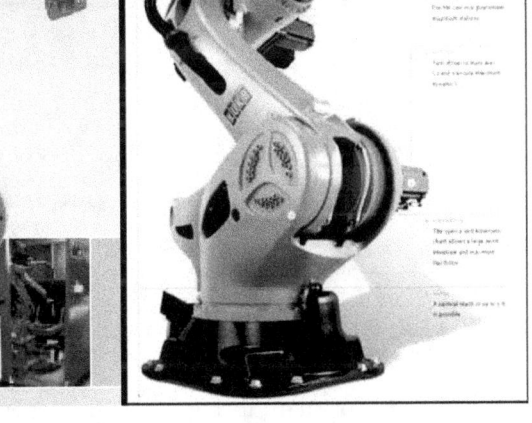
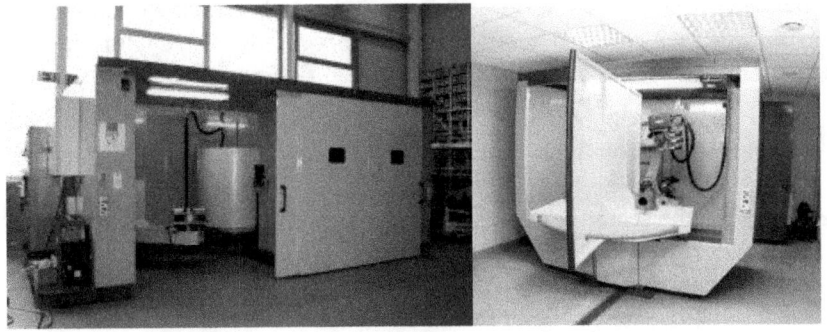

Fig. 3. *Miscellaneous modern structures 6R (anthropomorphic)*

In Figure 4 is illustrated geometro-cinematic a structure of basic 3R.

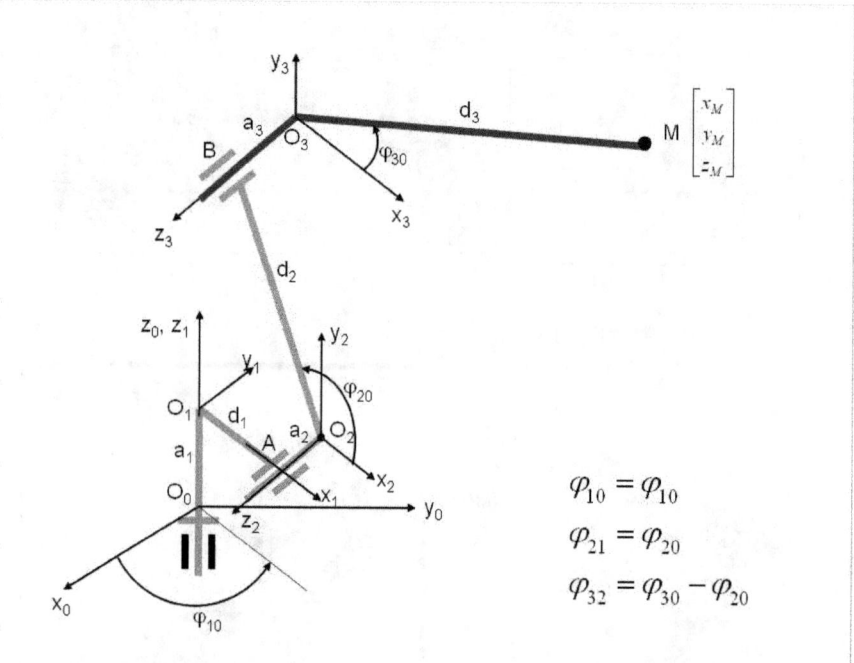

Fig. 4. *Schematic geometro-cinematic a structure 3R modern (anthropomorphic)*

From this platform can be studied by the addition any other scheme, modern, n-R.

Platform (system) in Fig. 4 has three degrees of freedom, made by three actuators (motors) or actuators. The first electric motor drives the whole system into a rotation around a vertical axis $O_0 z_0$. The motor (actuator) the number 1 is mounted on a fixed member (frame 0) and drives the movable element 1 in rotation around a vertical axis. On the mobile element 1, then all other elements are built (components of the system).

Follow a plan driveline (vertical), consisting of two mobile elements and two kinematic couplings engines. Its cinematic mobile elements 2 and 3, all being moved by the second actuator mounted in A couple, fixed on the element 1. So the second electric motor attached to the item 1 will move the item 2 (which result in rotation relative to the element 1), but it will automatically move the entire driveline 2-3.

The last (recent) actuator (electric motor) fixed to the element 2 in B, will rotate the element 3 (relative with respect 2).

The φ_{10} rotation (performed by the first actuator) is relative (between the elements 1 and 0) and absolute (between the elements 1 and 0).

The φ_{20} rotation (performed by the second actuator) is relative (between the elements 2 and 1) and absolute (between the elements 2 and 1).

The $\theta=\varphi_{32}$ rotation performed by the third actuator is only relative (between the elements 3 and 2), the corresponding absolute (between the elements 3 and 0) being a function of $\theta=\varphi_{32}$ and φ_{20}.

Driveline 2-3 (consisting of mobile cinematic elements 2 and 3) is a plan driveline falling into one plan or one or more parallel planes. It represents a unique cinematic that will be studied separately. We will consider the item 1 that caught powertrain 2-3 as fixed, the kinematic motor couplings A(O₂) și B(O₃) becoming the first fixed coupler and the second mobile coupler, both being kinematic couplings C₅ rotation.

In order to determine the degree of mobility of the kinematic chain plan 2-3, applies the structural formula given by the relation (1), where m is the number of moving parts of the kinematic chain level (in this case m=2; the case of the two kinematic elements, mobile, denoted by 2 and 3) and C₅ represent the number of kinematic couplings fifth grade (in the case of C₅=2; using couplers A and B or O₂ și O₃).

$$M_3 = 3\cdot m - 2\cdot C_5 = 3\cdot 2 - 2\cdot 2 = 6 - 4 = 2 \qquad (1)$$

The 2-3 kinematic chain, having the degree of mobility 2, to be driven by the two motors.

It is preferred that the two actuators are two electric motors, DC or alternating. But the drive can be achieved and with different engines: hydraulic motors, pneumatic, sonic, etc..

Structural diagram kinematic chain plan 2-3 (fig. 5) resembles with the cinematic scheme.

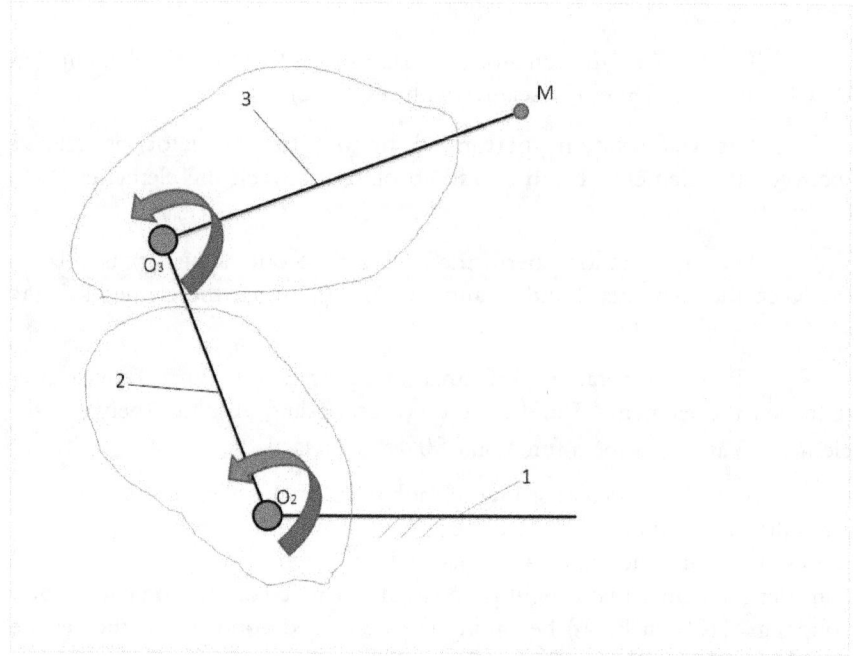

Fig. 5. *Structural Diagram kinematic chain plan 2-3 connected to item 1 considered fixed*

The conductor 2 is linked to the element considered fixed 1 through coupler O_2 engines, and the conductor 3 is linked to the mobile element 2 through coupler engines O_3.

Rezultă un lanț cinematic deschis cu două grade de mobilitate, realizate de cele două actuatoare, adică de cele două motoare electrice, montate în cuplele cinematice motoare A și B sau O_2 respectiv O_3.

This results in a kinematic chain open with two degree of mobility, which can be realized by the two actuators, i.e. the two electric motors, mounted in kinematic couplers engines A and B or O_2 respectively O_3.

Chapter 02_ The MP-3R Inverse Kinematics

1. Introduction

Although the anthropomorphic robots, have different structural forms, in recent years have been developed especially those with rotating movements, with three or more axis [1-27]. Constructive basis is represented by a robot with three degrees of freedom (a robot with three axes of rotation) [1]. If we study (analyze) an anthropomorphic robot with three axes of rotation (which represents the main movements, absolutely necessary), we already have a base system, on which we can then add other movements (secondary, additional). The base system has three rotary axes: a vertical axis (by this axis all the system is rotated, for positioning), and two horizontal axes (each making possible a rotation of an arm). Calculations were arranged and in the matrix form.

In direct kinematics, known kinematic parameters (input parameters) are absolute rotation angles of the three mobile elements: φ_{10}, φ_{20}, φ_{30}, the rotation angles of the three actuators (electric motors, mounted in the rotational kinematic couplings), and the determined parameters (output parameters) are the three absolute coordinates x_M, y_M, z_M of the point M, ie kinematic parameters (coordinates) of the endeffector (which can be a hand, to grabbed, a soldering tip, painted, cut, etc).

2. Geometry and Direct Kinematic, to the MP-3R

Kinematics of serial manipulators and robots will be illustrated by a 3R kinematic model (see Fig. 01), a medium difficulty system, ideal for understanding the phenomenon, but also to specify the basic knowledge necessary for starting calculations for systems simpler and more complex.

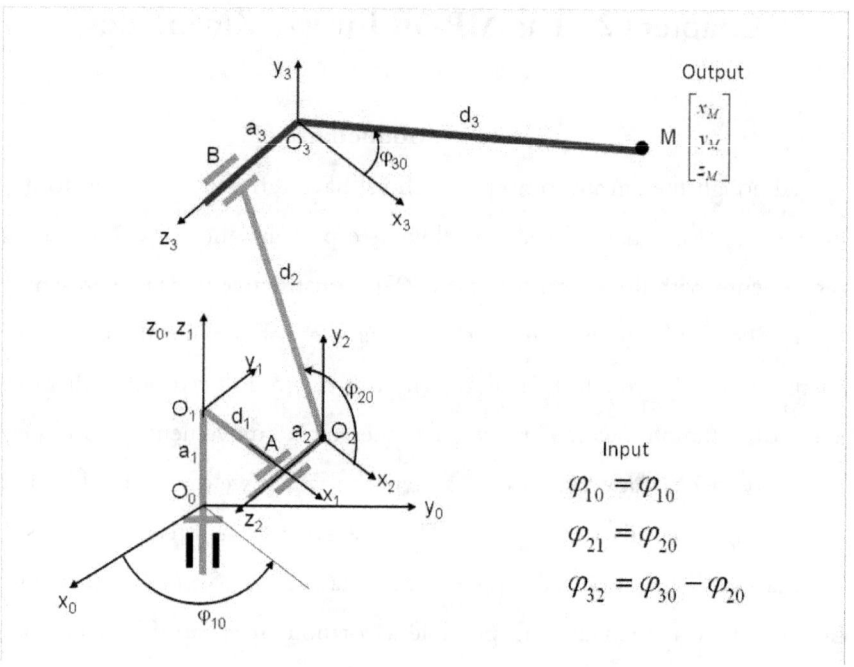

Fig. 1 *Geometry and direct kinematics to a MP-3R*

Fixed coordinate system was noted with $x_0 O_0 y_0 z_0$. Mobile systems related to (reinforced by) the three mobile elements (1, 2, 3) have indices 1, 2 and 3. Their orientation was chosen conveniently. Known kinematic parameters (input parameters in direct kinematics) are absolute rotation angles of the three mobile elements: φ_{10}, φ_{20}, φ_{30}, the rotation angles of the three actuators (electric motors, mounted in the rotational kinematic couplings). Determined parameters (output parameters) are the three absolute coordinates x_M, y_M, z_M of the point M, ie kinematic parameters (coordinates) of the endeffector (which can be a hand, to grabbed, a soldering tip, painted, cut, etc ...).

To begin we write vector matrix (A_{01}) which change the coordinates of the origin of the coordinate system, by linear moving (displacement) from

O_0 to O_1, when the axes remain parallel to each other permanently (see Eq. 2.1).

$$A_{01} = \begin{bmatrix} 0 \\ 0 \\ a_1 \end{bmatrix} \qquad (2.1)$$

Next we write the rotation matrix T_{01}, which rotates system $x_1O_1y_1z_1$ in rapport with the system $x_0O_0y_0z_0$ (it is a 3x3 square matrix; see the relationship 2.2).

$$T_{01} = \begin{bmatrix} \alpha_x & \beta_x & \gamma_x \\ \alpha_y & \beta_y & \gamma_y \\ \alpha_z & \beta_z & \gamma_z \end{bmatrix} = \begin{bmatrix} \cos\varphi_{10} & -\sin\varphi_{10} & 0 \\ \sin\varphi_{10} & \cos\varphi_{10} & 0 \\ 0 & 0 & 1 \end{bmatrix} \qquad (2.2)$$

On the first column (which represents the coordinates of the rotated axis O_1x_1) it writes the coordinates of the unit vector of O_1x_1 in rapport of the old system $x_0O_0y_0z_0$ (translated into O_1 but without rotation; see the relationship 2.3).

$$\begin{bmatrix} \alpha_x \\ \alpha_y \\ \alpha_z \end{bmatrix} \qquad (2.3)$$

On the second column of the matrix T_{01} it writes the coordinates of the unit vector of the rotated axis O_1y_1 in rapport of the old system $x_0O_0y_0z_0$ (translated into O_1 but without rotation system; see the relationship 2.4).

$$\begin{bmatrix} \beta_x \\ \beta_y \\ \beta_z \end{bmatrix} \qquad (2.4)$$

On the third column of the matrix T_{01} it writes the coordinates of the unit vector of the rotated axis O_1z_1 in rapport of the old system $x_0O_0y_0z_0$ (translated into O_1 but without rotation system; see the relationship 2.5).

$$\begin{bmatrix} \gamma_x \\ \gamma_y \\ \gamma_z \end{bmatrix} \quad (2.5)$$

In the elected case (figure 1), the unit vector of the rotated axis O_1x_1, has in rapport of the old system $x_0O_0y_0z_0$, translated into O_1 without rotation, the coordinates given by the column unit vector (relationship 2.6).

$$\begin{bmatrix} \alpha_x = 1 \cdot \cos\varphi_{10} = \cos\varphi_{10} \\ \alpha_y = 1 \cdot \sin\varphi_{10} = \sin\varphi_{10} \\ \alpha_z = 1 \cdot \cos 90^\circ = 1 \cdot 0 = 0 \end{bmatrix} \quad (2.6)$$

The unit vector of the rotated axis O_1y_1, has in rapport of the old system of axes $x_0O_0y_0z_0$ (translated into O1 without rotation), coordinates data unit vector column (relationship 2.7).

$$\begin{bmatrix} \beta_x = 1 \cdot \cos(\pi/2 + \varphi_{10}) = -\sin\varphi_{10} \\ \beta_y = 1 \cdot \sin(\pi/2 + \varphi_{10}) = \cos\varphi_{10} \\ \beta_z = 1 \cdot \cos(\pi/2) = 1 \cdot 0 = 0 \end{bmatrix} \quad (2.7)$$

The unit vector of the rotated axis O_1z_1 has in rapport of the old system of axes $x_0O_0y_0z_0$ (translated into O_1 without rotation), coordinates data unit vector column (relationship 2.8).

$$\begin{bmatrix} \gamma_x = 1 \cdot \cos 90^\circ = 1 \cdot 0 = 0 \\ \gamma_y = 1 \cdot \cos 90^\circ = 1 \cdot 0 = 0 \\ \gamma_z = 1 \cdot \cos 0^\circ = 1 \cdot 1 = 1 \end{bmatrix} \quad (2.8)$$

See the obtained matrix T_{01} (relationship 2.2).

Transition from the coordinate system $x_1O_1y_1z_1$ to the coordinate system $x_2O_2y_2z_2$ is done in two distinct phases. The first phase is a translation of the entire system so that (axes being parallel with them itself) the center O_1 to move into the center O_2; then the second stage in which it done the rotation of system of axes, and the center O remains fixed permanently.

The translation of the system from point 1 to the point 2 (see the relationship 2.9) is doing by the column vector, matrix A_{12}.

$$A_{12} = \begin{bmatrix} d_1 \\ a_2 \\ 0 \end{bmatrix} \qquad (2.9)$$

On the old O1x1 axis O2 has been moved with d1, on the old axis O1y1 O2 has been moved with a2, and on the old O1z1 axis O2 has not been moved.

The unit vector of the O_2x_2 axis has in rapport of the old system $x_1O_1y_1z_1$ (translated but not rotated) the next coordinates (expression 2.10).

$$\alpha_x = 1; \quad \alpha_y = 0; \quad \alpha_z = 0 \qquad (2.10)$$

The unit vector of the O_2y_2 axis has in rapport of the old system $x_1O_1y_1z_1$ (translated in O_2 but not rotated) the next coordinates (expression 2.11).

$$\beta_x = 0; \quad \beta_y = 0; \quad \beta_z = 1 \qquad (2.11)$$

The unit vector of the O_2z_2 axis has in rapport of the old system $x_1O_1y_1z_1$ (translated in O_2 but not rotated) the coordinates given by the expression 2.12.

$$\gamma_x = 0; \quad \gamma_y = -1; \quad \gamma_z = 0 \qquad (2.12)$$

The transfer square matrix (the rotation matrix: T12) is writing with relationship 2.13.

$$T_{12} = \begin{bmatrix} \alpha_x & \beta_x & \gamma_x \\ \alpha_y & \beta_y & \gamma_y \\ \alpha_z & \beta_z & \gamma_z \end{bmatrix} = \begin{bmatrix} 1 & 0 & 0 \\ 0 & 0 & -1 \\ 0 & 1 & 0 \end{bmatrix} \qquad (2.13)$$

Transition from the coordinate system $x_2O_2y_2z_2$ to the coordinate system $x_3O_3y_3z_3$ is done in two distinct phases. The first phase is a translation of the entire system so that (axes being parallel with them itself) the center O_2 to move into the center O_3; then the second stage in which it done the rotation of system of axes, and the center O_3 remains fixed permanently.

First O_2 is moving into O_3 (axes being parallel with them itself; see the relationship 2.14).

$$A_{23} = \begin{bmatrix} d_2 \cdot \cos\varphi_{20} \\ d_2 \cdot \sin\varphi_{20} \\ -a_3 \end{bmatrix} \qquad (2.14)$$

Then O_3 remains fixed, and the axes of coordinate system are rotating. The unit vector of the O_3x_3 axis has in rapport of the coordinate system $x_2O_2y_2z_2$ (translated in O_3 but not rotated) the α coordinates (see expression 2.15):

$$\alpha_x = 1; \quad \alpha_y = 0; \quad \alpha_z = 0 \qquad (2.15)$$

The unit vector of the O_3y_3 axis has in rapport of the coordinate system $x_2O_2y_2z_2$ (translated in O_3 but not rotated) the β coordinates (see relationship 2.16):

$$\beta_x = 0; \quad \beta_y = 1; \quad \beta_z = 0 \qquad (2.16)$$

The unit vector of the O₃z₃ axis has in rapport of the coordinate system x₂O₂y₂z₂ (translated in O₃ but not rotated) the γ coordinates (see relationship 2.17):

$$\gamma_x = 0; \quad \gamma_y = 0; \quad \gamma_z = 1 \qquad (2.17)$$

In the model from the figure 1 the system x₃O₃y₃z₃ has not been rotated in rapport of the system x₂O₂y₂z₂ (from 2 to 3 held just a translation). In this case the rotation matrix is the unit matrix (expression 2.18).

$$T_{23} = \begin{bmatrix} \alpha_x & \beta_x & \gamma_x \\ \alpha_y & \beta_y & \gamma_y \\ \alpha_z & \beta_z & \gamma_z \end{bmatrix} = \begin{bmatrix} 1 & 0 & 0 \\ 0 & 1 & 0 \\ 0 & 0 & 1 \end{bmatrix} \qquad (2.18)$$

The column vector matrix that positions the point M in the coordinate system x₃O₃y₃z₃ is written with relation 2.19.

$$X_{3M} = \begin{bmatrix} x_{3M} \\ y_{3M} \\ z_{3M} \end{bmatrix} = \begin{bmatrix} d_3 \cdot \cos\varphi_{30} \\ d_3 \cdot \sin\varphi_{30} \\ 0 \end{bmatrix} \qquad (2.19)$$

Coordinates of the point M in the system (2) x₂O₂y₂z₂ are obtained by a transformation matrix which is having the form (2.20):

$$X_{2M} = A_{23} + T_{23} \cdot X_{3M} \qquad (2.20)$$

First, is performed the matrix product (relations 2.21):

$$T_{23} \cdot X_{3M} = \begin{bmatrix} 1 & 0 & 0 \\ 0 & 1 & 0 \\ 0 & 0 & 1 \end{bmatrix} \cdot \begin{bmatrix} d_3 \cdot \cos\varphi_{30} \\ d_3 \cdot \sin\varphi_{30} \\ 0 \end{bmatrix} = \begin{bmatrix} d_3 \cdot \cos\varphi_{30} \\ d_3 \cdot \sin\varphi_{30} \\ 0 \end{bmatrix} \quad (2.21)$$

Then, will be calculated X_{2M} (relationship 2.22).

$$X_{2M} = A_{23} + T_{23} \cdot X_{3M} = \begin{bmatrix} d_2 \cdot \cos\varphi_{20} \\ d_2 \cdot \sin\varphi_{20} \\ -a_3 \end{bmatrix} + \begin{bmatrix} d_3 \cdot \cos\varphi_{30} \\ d_3 \cdot \sin\varphi_{30} \\ 0 \end{bmatrix} =$$

$$= \begin{bmatrix} d_2 \cdot \cos\varphi_{20} + d_3 \cdot \cos\varphi_{30} \\ d_2 \cdot \sin\varphi_{20} + d_3 \cdot \sin\varphi_{30} \\ -a_3 \end{bmatrix} \quad (2.22)$$

Coordinates of the point M in the system (1) $x_1 O_1 y_1 z_1$ are obtained by the relationships (2.23-2.25).

$$X_{1M} = A_{12} + T_{12} \cdot X_{2M} \quad (2.23)$$

$$T_{12} \cdot X_{2M} = \begin{bmatrix} 1 & 0 & 0 \\ 0 & 0 & -1 \\ 0 & 1 & 0 \end{bmatrix} \cdot \begin{bmatrix} d_2 \cdot \cos\varphi_{20} + d_3 \cdot \cos\varphi_{30} \\ d_2 \cdot \sin\varphi_{20} + d_3 \cdot \sin\varphi_{30} \\ -a_3 \end{bmatrix} =$$

$$= \begin{bmatrix} d_2 \cdot \cos\varphi_{20} + d_3 \cdot \cos\varphi_{30} \\ a_3 \\ d_2 \cdot \sin\varphi_{20} + d_3 \cdot \sin\varphi_{30} \end{bmatrix} \quad (2.24)$$

$$X_{1M} = A_{12} + T_{12} \cdot X_{2M} = \begin{bmatrix} d_1 \\ a_2 \\ 0 \end{bmatrix} + \begin{bmatrix} d_2 \cdot \cos\varphi_{20} + d_3 \cdot \cos\varphi_{30} \\ a_3 \\ d_2 \cdot \sin\varphi_{20} + d_3 \cdot \sin\varphi_{30} \end{bmatrix} =$$

$$= \begin{bmatrix} d_1 + d_2 \cdot \cos\varphi_{20} + d_3 \cdot \cos\varphi_{30} \\ a_2 + a_3 \\ d_2 \cdot \sin\varphi_{20} + d_3 \cdot \sin\varphi_{30} \end{bmatrix} \quad (2.25)$$

Coordinates of the point M in the fixed system $x_0O_0y_0z_0$, are written with the relationships (2.26-2.27, 2.27', 2.28).

$$X_{0M} = A_{01} + T_{01} \cdot X_{1M} \quad (2.26)$$

$$T_{01} \cdot X_{1M} = \begin{bmatrix} \cos\varphi_{10} & -\sin\varphi_{10} & 0 \\ \sin\varphi_{10} & \cos\varphi_{10} & 0 \\ 0 & 0 & 1 \end{bmatrix} \cdot \begin{bmatrix} d_1 + d_2 \cdot \cos\varphi_{20} + d_3 \cdot \cos\varphi_{30} \\ a_2 + a_3 \\ d_2 \cdot \sin\varphi_{20} + d_3 \cdot \sin\varphi_{30} \end{bmatrix} \quad (2.27)$$

$$T_{01} \cdot X_{1M} = \begin{bmatrix} (d_1 + d_2 \cdot \cos\varphi_{20} + d_3 \cdot \cos\varphi_{30}) \cdot \cos\varphi_{10} - (a_2 + a_3) \cdot \sin\varphi_{10} \\ (d_1 + d_2 \cdot \cos\varphi_{20} + d_3 \cdot \cos\varphi_{30}) \cdot \sin\varphi_{10} + (a_2 + a_3) \cdot \cos\varphi_{10} \\ d_2 \cdot \sin\varphi_{20} + d_3 \cdot \sin\varphi_{30} \end{bmatrix} \quad (2.27')$$

$$X_{0M} = A_{01} + T_{01} \cdot X_{1M} =$$
$$= \begin{bmatrix} 0 \\ 0 \\ a_1 \end{bmatrix} + \begin{bmatrix} (d_1 + d_2 \cdot \cos\varphi_{20} + d_3 \cdot \cos\varphi_{30}) \cdot \cos\varphi_{10} - (a_2 + a_3) \cdot \sin\varphi_{10} \\ (d_1 + d_2 \cdot \cos\varphi_{20} + d_3 \cdot \cos\varphi_{30}) \cdot \sin\varphi_{10} + (a_2 + a_3) \cdot \cos\varphi_{10} \\ d_2 \cdot \sin\varphi_{20} + d_3 \cdot \sin\varphi_{30} \end{bmatrix} = \quad (2.28)$$
$$= \begin{bmatrix} (d_1 + d_2 \cdot \cos\varphi_{20} + d_3 \cdot \cos\varphi_{30}) \cdot \cos\varphi_{10} - (a_2 + a_3) \cdot \sin\varphi_{10} \\ (d_1 + d_2 \cdot \cos\varphi_{20} + d_3 \cdot \cos\varphi_{30}) \cdot \sin\varphi_{10} + (a_2 + a_3) \cdot \cos\varphi_{10} \\ a_1 + d_2 \cdot \sin\varphi_{20} + d_3 \cdot \sin\varphi_{30} \end{bmatrix}$$

X_{0M} is put in the form (2.29).

$$X_{0M} = \begin{bmatrix} x_M \\ y_M \\ z_M \end{bmatrix} =$$

$$\begin{bmatrix} d_1 \cdot \cos\varphi_{10} - a_2 \cdot \sin\varphi_{10} + d_2 \cdot \cos\varphi_{20} \cdot \cos\varphi_{10} - a_3 \cdot \sin\varphi_{10} + d_3 \cdot \cos\varphi_{30} \cdot \cos\varphi_{10} \\ d_1 \cdot \sin\varphi_{10} + a_2 \cdot \cos\varphi_{10} + d_2 \cdot \cos\varphi_{20} \cdot \sin\varphi_{10} + a_3 \cdot \cos\varphi_{10} + d_3 \cdot \cos\varphi_{30} \cdot \sin\varphi_{10} \\ a_1 + d_2 \cdot \sin\varphi_{20} + d_3 \cdot \sin\varphi_{30} \end{bmatrix}$$ (2.29)

The same calculations will be presented now by a direct method (having in view the matrix calculations 2.30).

$$\begin{aligned} X_{0M} &= A_{01} + T_{01} \cdot X_{1M} = A_{01} + T_{01} \cdot (A_{12} + T_{12} \cdot X_{2M}) = \\ &= A_{01} + T_{01} \cdot A_{12} + T_{01} \cdot T_{12} \cdot X_{2M} = \\ &= A_{01} + T_{01} \cdot A_{12} + T_{01} \cdot T_{12} \cdot (A_{23} + T_{23} \cdot X_{3M}) = \\ &= A_{01} + T_{01} \cdot A_{12} + T_{01} \cdot T_{12} \cdot A_{23} + T_{01} \cdot T_{12} \cdot T_{23} \cdot X_{3M} \end{aligned}$$ (2.30)

It keeps the relationship (2.30').

$$X_{0M} = A_{01} + T_{01} \cdot A_{12} + T_{01} \cdot T_{12} \cdot A_{23} + T_{01} \cdot T_{12} \cdot T_{23} \cdot X_{3M}$$ (2.30')

Now, we perform the matrix multiplications from expression 2.30' (relationships 2.31-2.35).

$$T_{01} \cdot A_{12} = \begin{bmatrix} \cos\varphi_{10} & -\sin\varphi_{10} & 0 \\ \sin\varphi_{10} & \cos\varphi_{10} & 0 \\ 0 & 0 & 1 \end{bmatrix} \cdot \begin{bmatrix} d_1 \\ a_2 \\ 0 \end{bmatrix} = \begin{bmatrix} d_1 \cdot \cos\varphi_{10} - a_2 \cdot \sin\varphi_{10} \\ d_1 \cdot \sin\varphi_{10} + a_2 \cdot \cos\varphi_{10} \\ 0 \end{bmatrix}$$ (2.31)

$$T_{01} \cdot T_{12} = \begin{bmatrix} \cos\varphi_{10} & -\sin\varphi_{10} & 0 \\ \sin\varphi_{10} & \cos\varphi_{10} & 0 \\ 0 & 0 & 1 \end{bmatrix} \cdot \begin{bmatrix} 1 & 0 & 0 \\ 0 & 0 & -1 \\ 0 & 1 & 0 \end{bmatrix} = \begin{bmatrix} \cos\varphi_{10} & 0 & \sin\varphi_{10} \\ \sin\varphi_{10} & 0 & -\cos\varphi_{10} \\ 0 & 1 & 0 \end{bmatrix}$$ (2.32)

$$T_{01} \cdot T_{12} \cdot A_{23} = \begin{bmatrix} \cos\varphi_{10} & 0 & \sin\varphi_{10} \\ \sin\varphi_{10} & 0 & -\cos\varphi_{10} \\ 0 & 1 & 0 \end{bmatrix} \cdot \begin{bmatrix} d_2 \cdot \cos\varphi_{20} \\ d_2 \cdot \sin\varphi_{20} \\ -a_3 \end{bmatrix} =$$
$$= \begin{bmatrix} d_2 \cdot \cos\varphi_{10} \cdot \cos\varphi_{20} - a_3 \cdot \sin\varphi_{10} \\ d_2 \cdot \sin\varphi_{10} \cdot \cos\varphi_{20} + a_3 \cdot \cos\varphi_{10} \\ d_2 \cdot \sin\varphi_{20} \end{bmatrix} \quad (2.33)$$

$$T_{01} \cdot T_{12} \cdot T_{23} = \begin{bmatrix} \cos\varphi_{10} & 0 & \sin\varphi_{10} \\ \sin\varphi_{10} & 0 & -\cos\varphi_{10} \\ 0 & 1 & 0 \end{bmatrix} \cdot \begin{bmatrix} 1 & 0 & 0 \\ 0 & 1 & 0 \\ 0 & 0 & 1 \end{bmatrix} =$$
$$= \begin{bmatrix} \cos\varphi_{10} & 0 & \sin\varphi_{10} \\ \sin\varphi_{10} & 0 & -\cos\varphi_{10} \\ 0 & 1 & 0 \end{bmatrix} \quad (2.34)$$

$$T_{01} \cdot T_{12} \cdot T_{23} \cdot X_{3M} = \begin{bmatrix} \cos\varphi_{10} & 0 & \sin\varphi_{10} \\ \sin\varphi_{10} & 0 & -\cos\varphi_{10} \\ 0 & 1 & 0 \end{bmatrix} \cdot \begin{bmatrix} d_3 \cdot \cos\varphi_{30} \\ d_3 \cdot \sin\varphi_{30} \\ 0 \end{bmatrix} =$$
$$= \begin{bmatrix} d_3 \cdot \cos\varphi_{10} \cdot \cos\varphi_{30} \\ d_3 \cdot \sin\varphi_{10} \cdot \cos\varphi_{30} \\ d_3 \cdot \sin\varphi_{30} \end{bmatrix} \quad (2.35)$$

The expression (2.30') takes the form (2.36).

$$X_{OM} = \begin{bmatrix} 0 \\ 0 \\ a_1 \end{bmatrix} + \begin{bmatrix} d_1 \cdot \cos\varphi_{10} - a_2 \cdot \sin\varphi_{10} \\ d_1 \cdot \sin\varphi_{10} + a_2 \cdot \cos\varphi_{10} \\ 0 \end{bmatrix} + \begin{bmatrix} d_2 \cdot \cos\varphi_{10} \cdot \cos\varphi_{20} - a_3 \cdot \sin\varphi_{10} \\ d_2 \cdot \sin\varphi_{10} \cdot \cos\varphi_{20} + a_3 \cdot \cos\varphi_{10} \\ d_2 \cdot \sin\varphi_{20} \end{bmatrix} +$$

$$+ \begin{bmatrix} d_3 \cdot \cos\varphi_{10} \cdot \cos\varphi_{30} \\ d_3 \cdot \sin\varphi_{10} \cdot \cos\varphi_{30} \\ d_3 \cdot \sin\varphi_{30} \end{bmatrix} = \begin{bmatrix} x_M \\ y_M \\ z_M \end{bmatrix} =$$

$$= \begin{bmatrix} d_1 \cdot \cos\varphi_{10} - a_2 \cdot \sin\varphi_{10} + d_2 \cdot \cos\varphi_{20} \cdot \cos\varphi_{10} - a_3 \cdot \sin\varphi_{10} + d_3 \cdot \cos\varphi_{30} \cdot \cos\varphi_{10} \\ d_1 \cdot \sin\varphi_{10} + a_2 \cdot \cos\varphi_{10} + d_2 \cdot \cos\varphi_{20} \cdot \sin\varphi_{10} + a_3 \cdot \cos\varphi_{10} + d_3 \cdot \cos\varphi_{30} \cdot \sin\varphi_{10} \\ a_1 + d_2 \cdot \sin\varphi_{20} + d_3 \cdot \sin\varphi_{30} \end{bmatrix}$$

(2.36)

By the direct kinematics is obtained Cartesian coordinates x_M, y_M, z_M of the point M (the endeffector) in rapport with the three independent angular displacements φ_{10}, φ_{20}, φ_{30}, obtained using actuators (relationships 2.37-2.38).

$$\begin{cases} x_M = f_x(\varphi_{10}, \varphi_{20}, \varphi_{30}) \\ y_M = f_y(\varphi_{10}, \varphi_{20}, \varphi_{30}) \\ z_M = f_z(\varphi_{10}, \varphi_{20}, \varphi_{30}) \end{cases} \qquad (2.37)$$

$$\begin{cases} x_M = d_1 \cdot \cos\varphi_{10} - a_2 \cdot \sin\varphi_{10} + d_2 \cdot \cos\varphi_{20} \cdot \cos\varphi_{10} - \\ \quad - a_3 \cdot \sin\varphi_{10} + d_3 \cdot \cos\varphi_{30} \cdot \cos\varphi_{10} \\ y_M = d_1 \cdot \sin\varphi_{10} + a_2 \cdot \cos\varphi_{10} + d_2 \cdot \cos\varphi_{20} \cdot \sin\varphi_{10} + \\ \quad + a_3 \cdot \cos\varphi_{10} + d_3 \cdot \cos\varphi_{30} \cdot \sin\varphi_{10} \\ z_M = a_1 + d_2 \cdot \sin\varphi_{20} + d_3 \cdot \sin\varphi_{30} \end{cases} \qquad (2.38)$$

Calculations are performed with absolute angular movements (φ_{10}, φ_{20}, φ_{30}), but the actuators movements do not match (all) with the independent angular movements. They are determined as follows (expressions 2.39):

$$\begin{cases} \varphi_{10} = \varphi_{10} \\ \varphi_{21} = \varphi_{20} \\ \varphi_{32} = \varphi_{30} - \varphi_{20} \end{cases} \quad (2.39)$$

The first two actuators relative rotations coincide with the independent rotations (used in calculations), but the third actuator relative rotation is obtained as a difference between two absolute rotations (expressions 2.39). The velocities and the accelerations are obtained by the derivatives of the positions expressions (2.38) in rapport of the time.

3. The inverse geometry and inverse kinematics at a MP-3R

The inverse kinematic [2-8] at the serial robots and systems will be exemplified for the 3R kinematic model (see the Fig. 2).

In inverse kinematics, we already know the direct link relationships (3.1), and must determine the inverse relationships, ie to determine the independent rotations φ_{10}, φ_{20}, φ_{30} of the three mobile elements, based on kinematic parameters imposed to the endeffector x_M, y_M, z_M, known (forced).

With the independent determined angles, is then to be calculated the relative rotation movements, of the three driving motors, from the rotating couplers [7].

$$\begin{cases} x_M = d_3 \cos\varphi_{10} \cdot \cos\varphi_{30} + d_2 \cos\varphi_{10} \cdot \cos\varphi_{20} - \\ \quad - a_3 \sin\varphi_{10} + d_1 \cos\varphi_{10} - a_2 \sin\varphi_{10} \\ y_M = d_3 \sin\varphi_{10} \cdot \cos\varphi_{30} + d_2 \sin\varphi_{10} \cdot \cos\varphi_{20} + \\ \quad + a_3 \cos\varphi_{10} + d_1 \sin\varphi_{10} + a_2 \cos\varphi_{10} \\ z_M = d_3 \sin\varphi_{30} + d_2 \sin\varphi_{20} + a_1 \end{cases} \quad (3.1)$$

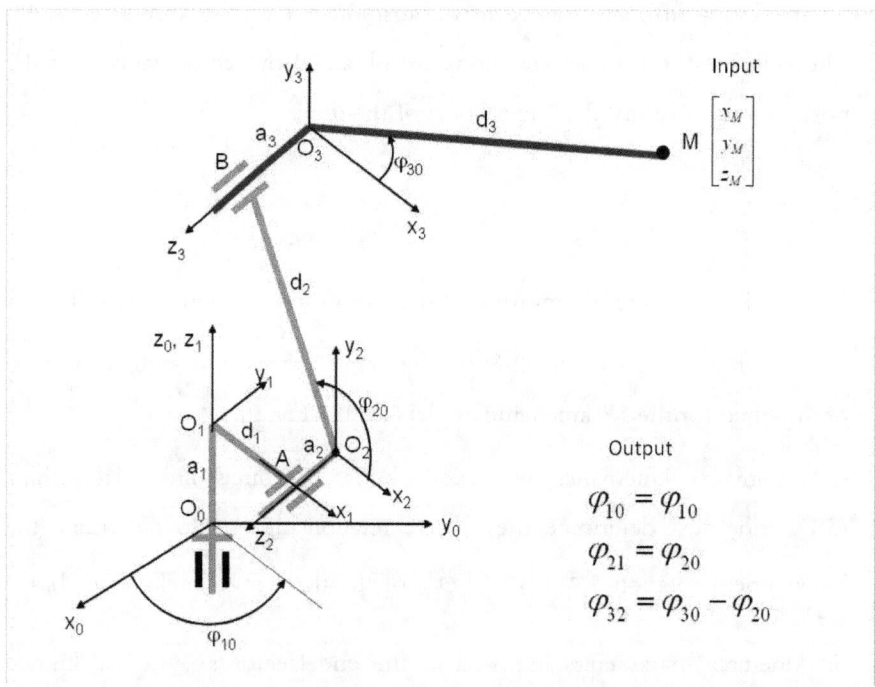

Fig. 2 *The inverse kinematic at the serial robots and systems, exemplified for the 3R kinematic model*

Fixed coordinate system was noted with x₀O₀y₀z₀. Mobile systems related to the three mobile elements (1, 2, 3) have the indices 1, 2 and 3. Their orientation was chosen conveniently.

System (3.1) is a system of three nonlinear equations (1.1-1.3) with three unknowns (φ_{10}, φ_{20}, φ_{30}) that must be determined; the system 3.1 equations, are rearranging in form that can be seen in the system (3.1 ').

$$\begin{cases} x_M = d_1 \cdot \cos\varphi_{10} - a_2 \cdot \sin\varphi_{10} + d_2 \cdot \cos\varphi_{20} \cdot \cos\varphi_{10} - \\ \quad - a_3 \cdot \sin\varphi_{10} + d_3 \cdot \cos\varphi_{30} \cdot \cos\varphi_{10} (1.1) \\ y_M = d_1 \cdot \sin\varphi_{10} + a_2 \cdot \cos\varphi_{10} + d_2 \cdot \cos\varphi_{20} \cdot \sin\varphi_{10} + \\ \quad + a_3 \cdot \cos\varphi_{10} + d_3 \cdot \cos\varphi_{30} \cdot \sin\varphi_{10} (1.2) \\ z_M = a_1 + d_2 \cdot \sin\varphi_{20} + d_3 \cdot \sin\varphi_{30} (1.3) \end{cases} \quad (3.1')$$

It aims to solve the system (3.1 ') directly, to obtain exact and independent solutions.

The first step is multiplying expression (1.1) with ($-\sin\varphi_{10}$) and relation (1.2) with ($\cos\varphi_{10}$); then is summing the two expressions and resulting trigonometric equation (3.2), which can be solved and gives the solutions (3.3-3.4).

$$-x_M \cdot \sin\varphi_{10} + y_M \cdot \cos\varphi_{10} = a_2 + a_3 \qquad (3.2)$$

$$\begin{cases} \cos\varphi_{10} = \dfrac{(a_2+a_3)\cdot y_M \pm x_M \cdot \sqrt{x_M^2+y_M^2-(a_2+a_3)^2}}{x_M^2+y_M^2} \\ \sin\varphi_{10} = \dfrac{-(a_2+a_3)\cdot x_M \pm y_M \cdot \sqrt{x_M^2+y_M^2-(a_2+a_3)^2}}{x_M^2+y_M^2} \end{cases} \quad (3.3)$$

One determines for the first independently parameter (φ_{10}), the trigonometric values of the functions cos and sin ($\cos\varphi_{10}$ and $\sin\varphi_{10}$), (system 3.3).

We can directly obtain an angle value, when we know sin and cos functions, using expression (3.4).

$$\varphi_{10} = \text{semn}(\sin\varphi_{10})\cdot \arccos(\cos\varphi_{10}) \quad (3.4)$$

Angle is given directly by the arccos function.

Sign of sinus (which can be +1 or -1) send the angle in its quadrant, in the top semicircle or the bottom.

The next step is multiplying expression (1.1) with ($\cos\varphi_{10}$) and relation (1.2) with ($\sin\varphi_{10}$); the two resulted are summed, and one obtains the trigonometric equation (3.5).

$$x_M \cdot \cos\varphi_{10} + y_M \cdot \sin\varphi_{10} - d_1 = d_2 \cdot \cos\varphi_{20} + d_3 \cdot \cos\varphi_{30} \quad (3.5)$$

This relation (3.5) together with the relation (1.3) forms the system (3.6), which generates the independent parameters (φ_{20} *and* φ_{30}, the last).

$$\begin{cases} x_M \cdot \cos\varphi_{10} + y_M \cdot \sin\varphi_{10} - d_1 = d_2 \cdot \cos\varphi_{20} + d_3 \cdot \cos\varphi_{30} \quad (3.5) \\ z_M - a_1 = d_2 \cdot \sin\varphi_{20} + d_3 \cdot \sin\varphi_{30} \quad (1.3) \end{cases} \quad (3.6)$$

With notations (3.7) one obtains for the equations system (3.6) the direct and exact solutions (3.8).

The equations (3.6) take the form (3.6').

$$\begin{cases} C_1 = d_2 \cdot \cos\varphi_{20} + d_3 \cdot \cos\varphi_{30} \quad (3.5') \\ C_2 = d_2 \cdot \sin\varphi_{20} + d_3 \cdot \sin\varphi_{30} \quad (1.3') \end{cases} \quad (3.6')$$

System (3.6') can be written in the form (3.6'').

$$\begin{cases} C_1 - d_2 \cdot \cos\varphi_{20} = d_3 \cdot \cos\varphi_{30} \quad (3.5'') \\ C_2 - d_2 \cdot \sin\varphi_{20} = d_3 \cdot \sin\varphi_{30} \quad (1.3'') \end{cases} \quad (3.6'')$$

Equations (3.6''), squared and added together, give the expression (3.6''').

$$K - 2 \cdot C_1 \cdot d_2 \cdot \cos\varphi_{20} = 2 \cdot C_2 \cdot d_2 \cdot \sin\varphi_{20} \quad (3.6''')$$

$$\begin{cases} C_1 = x_M \cdot \cos\varphi_{10} + y_M \cdot \sin\varphi_{10} - d_1 \\ C_2 = z_M - a_1 \\ k = C_1^2 + C_2^2 + d_2^2 - d_3^2 \end{cases} \quad (3.7)$$

From equation (3.6''') one obtains $\cos\varphi_{20}$, $\sin\varphi_{20}$, and φ_{20} (first relations of system 3.8), and using expressions (3.6'') it determines then $\cos\varphi_{30}$, $\sin\varphi_{30}$, and φ_{30} (last relations of system 3.8).

$$\begin{cases} \cos\varphi_{20} = \dfrac{k \cdot C_1 \pm C_2 \cdot \sqrt{4 \cdot C_1^2 \cdot d_2^2 + 4 \cdot C_2^2 \cdot d_2^2 - k^2}}{2 \cdot (C_1^2 + C_2^2) \cdot d_2} \\ \sin\varphi_{20} = \dfrac{k \cdot C_2 \mp C_1 \cdot \sqrt{4 \cdot C_1^2 \cdot d_2^2 + 4 \cdot C_2^2 \cdot d_2^2 - k^2}}{2 \cdot (C_1^2 + C_2^2) \cdot d_2} \\ \varphi_{20} = semn(\sin\varphi_{20}) \cdot \arccos(\cos\varphi_{20}) \\ \cos\varphi_{30} = \dfrac{C_1 - d_2 \cdot \cos\varphi_{20}}{d_3} \\ \sin\varphi_{30} = \dfrac{C_2 - d_2 \cdot \sin\varphi_{20}}{d_3} \\ \varphi_{30} = semn(\sin\varphi_{30}) \cdot \arccos(\cos\varphi_{30}) \end{cases} \qquad (3.8)$$

Conclusions

Kinematics of serial manipulators and robots can be illustrated by a 3R kinematic model, a medium difficulty system, ideal for understanding the phenomenon, but also to specify the basic knowledge necessary for starting calculations for systems simpler and more complex.

The chapter presents an original geometrical and kinematic method for the study of geometry and determining positions of a MP-3R structure. It presents shortly the MP-3R direct and inverse kinematics, the inverse kinematics being solved by an original exactly method. One presents shortly an original method to solve the robot inverse kinematics exemplified at the 3R-Robots (MP-3R).

If we study (analyze) an anthropomorphic robot with three axes of rotation (which represents the main movements, absolutely necessary), we already have a base system, on which we can then add other movements (secondary, additional). Calculations were arranged and in the matrix form.

References

1. Antonescu P., Mecanisme și manipulatoare, Editura Printech, Bucharest, 2000, p. 103-104.
2. Angeles J., s.a., An algorithm for inverse dynamics of n-axis general manipulator using Kane's equations, Computers Math. Applic, Vol.17, No.12, 1989.
3. Borrel P., Liegeois A., A study of manipulator inverse kinematic solutions with application to trajectory planning and workspace determination. In Prod. IEEE Int. Conf. Rob. and Aut., pp. 1180-1185, 1986.
4. Do W.Q.D., Yang, D.C.H. (1988). Inverse dynamic analysis and simulation of a platform type of robot. Journal of Robotic Systems, 5(3), p. 209-227.
5. Guglielmetti, P., Longchamp, R., A Closed Form Inverse Dynamics Model of the DELTA Parallel Robot, Symposium on Robot Control, Capri, Italia, 1994, p. 51-56.
6. Hollerbach J.M., Wrist-partitioned inverse kinematic accelerations and manipulator dynamics, International Journal of Robotic Research 2, 61-76 (1983).
7. Petrescu F.I., Grecu B., Comănescu Adr., Petrescu R.V., Some Mechanical Design Elements, Proceedings of International Conference Computational Mechanics and Virtual Engineering, COMEC 2009, October 2009, Brașov, Romania, pp. 520-525.
8. Powell I.L., B.A.Miere, The kinematic analysis and simulation of the parallel topology manipulator, The Marconi Review, 1982.
9. Raghavan, M., Roth, B., Solving polynomial systems for the kinematics analysis of mechanisms and robot manipulators, ASME J. of Mechanical Design, 117 (2), 1995, p.71-79.
10. Reboulet, C., Pigeyre, R., Hybrid Control of a 6 d.o.f. in parallel actuated micromanipulator mounted on a SCARA robot, Int J. of Robotics and Automation, 7 (1), 1992, p. 10-14.
11. Reddy M., a.o., Precise Non Linear Modeling of Flexible Link Flexible Joint Manipulator, IREMOS, Vol. 5, N. 3, June (Part B) 2012, p. 1368-1374.
12. Riesler, H., Zur Berechnung geschlossener Lösungen des inversen kinematischen Problems, Fortschritte der Robotik, 16, Vieweg, 1992.
13. Rong, H., Liang, C.,G., A Direct Displacement Solution to the Triangle- Platform 6-SPS Parallel Manipulator, 8th Congres on the Theory of Machines and Mechanisms, Prague, Cehoslovacia, 1991, p. 1237-1239.

14. Seeger G., Self-tuning of commercial manipulator based on an inverse dynamic model, J.Robotics Syst. 2 / 1990.
15. Sefrioui, J. and Gosselin, C.M., Étude et reprézentation des lieux de singularité des manipulateurs parallèles spheriques à trois degrés de liberté avec actionneurs prismatiques, in Mech. Mach. Theory Vol. 29, No.4, 1994, p. 559-579.
16. Shanmuga G., a.o., A Survey on Development of Inspection Robots: Kinematic Analysis, Workspace Simulation and Software Development, IREME, Vol. 6, N. 7, November 2012, p.1493-1507.
17. Smith S.T., Chetwynd D.G., Foundations of Ultraprecision Mechanism Design. Gordon and Breach Science Publishers, Switzerland, 1992.
18. Tadokorro, S., Control of parallel mechanisms. Advanced Robotics, 8 (6), 1994, p. 559-571.
19. Tahmasebi, F., Tsai, L-W., Jacobian and Stiffness Analysis of a Novel Class of Six-dof Parallel Minimanipulators, DE-Vol.47, Flexible Mechanisms, Dynamics and Analysis, ASME, 1992, p. 95-102.
20. Tsai L-W. Solving the inverse dynamics of a Stewart-Gough manipulator by the principle of virtual work. ASME J. of Mechanical Design, 122(1), Mars 2000, p. 3-9.
21. Vazquez, F., Marin, R., Trillo, J. L., Garrido, J., Object Oriented Modeling, Design & Simulation of Industrial Autonomous Mobile Robots, EURISCON, 1994, p. 361-371.
22. Walker, M., W., Orin, D.E., Efficient Dynamic Computer Simulation of Robotic Mechanisms, Journal of Dynamic Systems, Measurement and Control, vol 104; 1982, p 205-211.
23. Wampler, C,W., Forward displacement analysis of general six-in parallel SPS (Stewart) platform manipulators using some coordinates. Mechanism and Machine Theory, 31 (3), 1996, p. 331-337.
24. Wang J. et Gosselin C.M. A new approach for the dynamic analysis of parallel manipulators. Multibody System Dynamics, 2(3), Septembre 1998, p. 317-334.
25. Wu, Y., Gosselin, C., On the Synthesis on a Reactionless 6-DOF Parallel Mechanism using Planar Four-Bar Linkages, Proc. of the Workshop on Fundamentals Issues and Future Research Directions for Parallel mechanism and Manipulators, Canada, 2002, p. 310-316.
26. Yang, K-H., Park, Y-S., Dynamic Stability Analysis of a Flexible Four-Bar Mechanism and its Experimental Investigation, Mech. Mach. Theory, Vol. 33, No. 3, 1998, p. 307-320.
27. Zhang C., Song S-M., Forward Position Analysis of Nearly General Stewart Platforms, ASME Robotics, Spatial Mechanisms and Mechanical Systems, DE-Vol 15, 1992, p. 81-87.

Chapter 03_The Structure and Direct Kinematics

Introduction

Moving mechanical structures are used increasingly in almost all vital sectors of humanity (Tong, 2013). The robots are able to process integrated circuits sizes micro and nano, on which the man they can be seen even with electron microscopy (Lee, 2013). Dyeing parts in toxic environments, working in chemical and radioactive environments, or at depths and pressures at the bottom of huge oceans, or even cosmic space conquest and visiting exo-planets, are now possible, and were turned into from the dream in reality, because mechanical platforms sequential gearbox (Dong, 2013).

The man will be able to carry out its mission supreme, conqueror of new galaxies, because mechanical systems sequential gearbox (Perumaal, 2013).

Robots were developed and diversified, different aspects, but today, they start to be directed on two major categories: systems serial and parallel systems (Padula, 2013).

Parallel systems are more solid, but more difficult to designed and handled, which serial systems were those which have developed the most.

Serial systems and they have different constructive diagrams, but over the last 30 years have been channelled on anthropomorphic structures (Reddy, 2012).

These structures are made up of simple components and couplers for rotation.

Their great advantage is fast movements, good dynamics, a high accuracy, a construction of simple modules STAS, economy of materials, low cost, and high reliability.

One disadvantage of less accurate has been removed because of stepper motors.

Compared with parallel systems (more solid but more cumbersome) serial systems may pose, and the disadvantage of stability something lower.

This disadvantage begins to be exceeded today in smart mode, through the construction of serial systems made up of elements doubled (in parallel).

This last invention, will lead to the strengthening of serial systems, and to their consolidation like the indisputable leader in diversity of mechatronics and robotic systems.

This work starts from a main idea, to study these systems on a single model, 3R, which has finally main movements lying on a single plane model, 2R.

The structure of moving (serial) mechanical systems

The most commonly used serial structures over the last 20 or 30 years are those of type 3R, 4R, 5R, 6R, having as constituents essential basic kinematic chain 3R, robot antropomorf (RRR), where main rotation around a vertical axis, causes the construction (Petrescu, 2012).

There are then a basic kinematic chain which has two revolutions 'bokeh' (two actuators, i.e. two motors) who work permanently in one plane, and immediately after main support which supports and rotates vertically complete assembly (Petrescu, 2011).

This basic structure (Tang, 2013), 3R, a meet me at all robots serial manufactured on the principle of rotations. Vertical Bracket is was the same, but the drive train as follows, with the two turns situated in a plane can be positioned vertically (most often; the robots anthropomorphic, fig. 1b), or horizontally (case robots scale, fig. 1a).

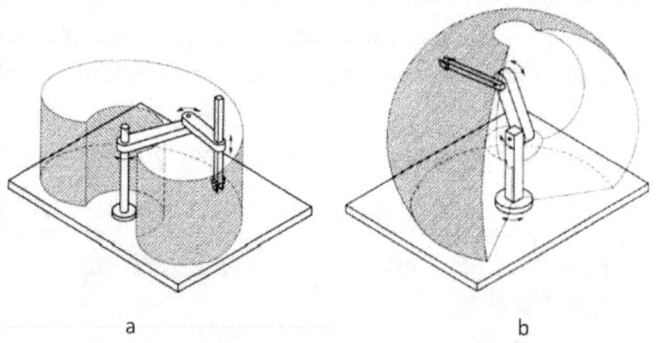

Figure 1. *The basic structures 3R (a-scale structure ; b-anthropomorphic structure)*

It can thus passes from the study spatial movement, which is more difficult, to the study motion plane, basic movement, for all the robots and fillers serial movements of rotation.

Moving flat, horizontal or vertical, shall be undertaken far more easily than the spatial integration with the convenience simple in the space of which it is part (Garcia, 2007).

We will exemplify the basic structure existing in a few serial platforms of rotation, these being the most generalized (more widespread) at the present time. On this basic model (3R) have developed further robots 6R (He, 2013) today (fig. 2, rely only on revolutions using actuator that only electric motors, compact); they have a hardness greater penetration while maintaining the flexibility and models 3R, 4R and 5R.

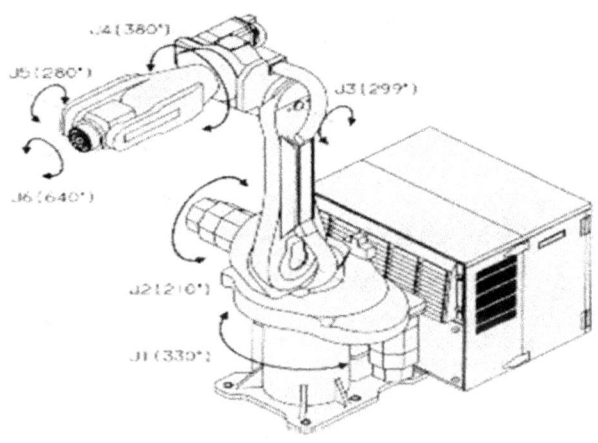

Figure 2. *Structure 6R (anthropomorphic structure)*

Almost all major companies come today with models 6R (which they improve continuously).

Why they have imposed today these models of robots (after tens of years of diversity was the word of order); may and of the need for standardisation, or to find a common solution, after a huge portion (however are not yet the only robots use of category serial, but they also have the widest spread).

The six turns (full elimination of translations, who bring many disadvantages due to coupler T itself) are operating easier, faster, with higher yield, more reliable, more compact and more secure; basic revolutions, remain all first three, the other three turns (additional) having the role of position may well end device, the endeffector. Results and that the baseline study (required) it is still not for a 3R.

This can be seen in the latest models of the various firms producing robots (fig. 3, Kawasaki, Romat, Fanuc, Motoman, Kuka, etc). And the structures used inside cells sequential gearbox are constructed generally in a similar way.

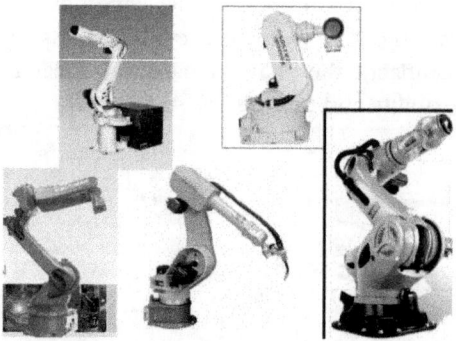

Figure 3. *Miscellaneous modern structures 6R (anthropomorphic)*

In figure 4 is illustrated geometro-cinematic a structure of basic 3R.

Starting from this platform may be studied by addition any other scheme, n-R modern.

The platform (system) in figure 4, has three degree of mobility, which can be realized by three actuators (electric motors). First electric motor drives your whole system in a rotating around a vertical spindle $O_0 z_0$. Engine (actuator) number 1, is mounted on the fixed (frame, 0) and causes mobile element 1 in a rotating around a vertical axis. The mobile element 1, then all the other elements (components) of the system.

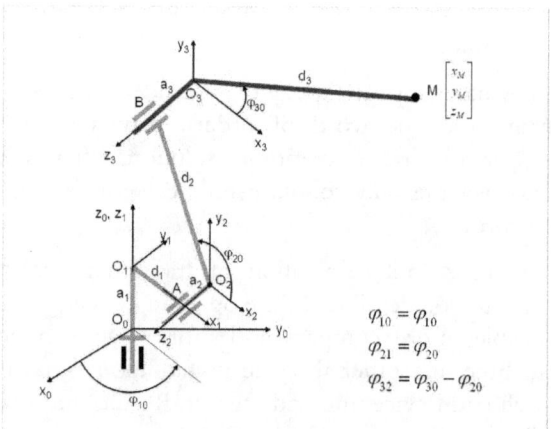

Figure 4. *Layout geometro-cinematic a modern structures 3R (anthropomorphic)*

Follows a kinematic chain plane (vertical), composed of two moving components and two couplers 'bokeh' engines. It's about the kinematic elements mobile 2 and 3, the assembly 2-3 being moved by the actuator of the second mounted to engage A (O_2), fixed on item 1. Therefore the second electric motor attached to the component 1 will result in the item 2 of rotating relative to the item 1, but he will move automatically entire drive train 2-3.

Last actuator (electric motor) attached to the item (2), and in B (O_3), will rotate item 3 (relative in relation to 2).

Rotation φ_{10} carried out by the first actuator, is and relative (between items 1 and 0) and absolute (between elements 1 and 0).

Rotation φ_{20} carried out by the second actuator, is and relative (between items 2 and 1) and absolute (between items 2 and 0), due to the arrangement.

Rotation $\theta=\varphi_{32}$ carried out by the third actuator, is only relative (between items 3 and 2); the corresponding absolute (between items 3 and 0) as a function of $\theta=\varphi_{32}$ and φ_{20}.

The drive train 2-3 (consisting of kinematic elements 2 and 3) is a kinematic chain plan, which fall within a single plane or in one or more of the other plane parallel to each other. It is a special kinematic system, which will be examined separately. It shall be considered as item 1 of which is caught the drive train 2-3 as being fixed, couplers kinematic engines A(O_2) and B(O_3) becoming first fixed coupler, and the coupler to make two mobile, which are both couplers kinematic C5, of rotation.

Determination of the extent of mobility of kinematic chain plan 2-3, apply structural formula given by the relationship (1), where m represents the number of elements of the kinematic chain mobile plan, in our case m=2 (as we are talking about the two kinematic elements element noted with 2 and 3), and C_5 represents the number of couplers 'bokeh' fifth-class, in the case in point $C_5=2$ (in the case of couplers A and B or O_2 and O_3).

$$M_3 = 3 \cdot m - 2 \cdot C_5 = 3 \cdot 2 - 2 \cdot 2 = 6 - 4 = 2 \qquad (1)$$

The drive train 2-3 having degree of mobility 2, must be actuated by two motors (Liu, 2013).

It is preferred that the two actuators to be two electric motors, a direct current, or alternately. Action can be done but with different engines. Hydraulic motors, pneumatic, sonic, etc.

Structural schematic kinematic chain plan 2-3 (fig. 5) resembles with cinematic scheme.

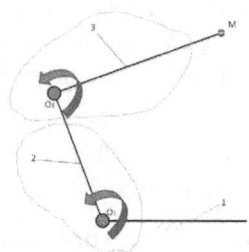

Figure 5. *Structural schematic kinematic chain plan 2-3 linked to item 1 considered fixed*

The conductor 2 is linked to the element considered fixed coupler 1 by the term O_2 engines, and drive element 3 is connected to the element mobile 2 by engage engines O_3. This results in a kinematic chain open with two degree of mobility, which can be realized by the two actuators, i.e. the two electric motors, mounted on 'bokeh' couplers engines A and B or O_2 or O_3 (Garcia-Murillo, 2013).

The kinematics chain direct plan 2-3

In figure 6 can be monitored kinematic chain schematic plan 2-3 opened.

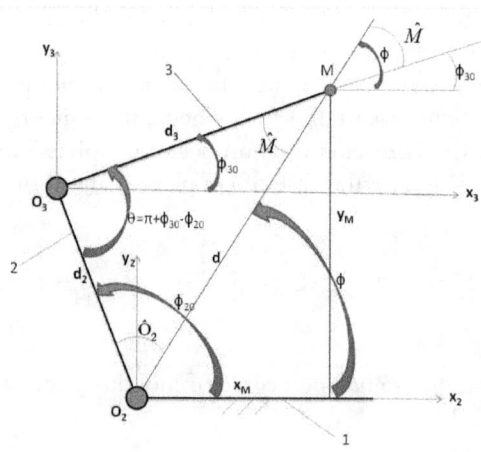

Figure 6. *Cinematic schematic kinematic chain plan 2-3 linked to item 1 considered fixed*

In the direct kinematics parameters (Wang, 2013) are known the parameters cinematic Φ20 and Φ30 and must be determined by analytical calculation parameters x_M and y_M, which represents the co-ordinates of scaling point M (endeffector M). Designing vectors $d_2 + d_3$ on cartesian axis system considered fixed, xOy, identical to $x_2O_2y_2$. Obtain the equation scaling (2) (Wen, 2012).

$$\begin{cases} x_{2M} \equiv x_M = x_{O_3} + x_{3M} = \\ = d_2 \cdot \cos\phi_{20} + d_3 \cdot \cos\phi_{30} = d \cdot \cos\phi \\ y_{2M} \equiv y_M = y_{O_3} + y_{3M} = \\ = d_2 \cdot \sin\phi_{20} + d_3 \cdot \sin\phi_{30} = d \cdot \sin\phi \end{cases} \quad (2)$$

After you determine the co-ordinates of Cartesian point M using data relations of the system (2), may be obtained immediately and parameters using angle relationships established within the framework of the system (3).

$$\begin{cases} d^2 = x_M^2 + y_M^2 ;\quad d = \sqrt{x_M^2 + y_M^2} \\ \cos\phi = \dfrac{x_M}{d} = \dfrac{x_M}{\sqrt{x_M^2 + y_M^2}} \\ \sin\phi = \dfrac{y_M}{d} = \dfrac{y_M}{\sqrt{x_M^2 + y_M^2}} \\ \phi = sign(\sin\phi) \cdot \arccos(\cos\phi) \end{cases} \quad (3)$$

The system (2) is being written brief in the form (4) which shall be derived as a function of time, resulting in the system (5), which derived with time generates in turn the accelerations (6).

$$\begin{cases} x_M = d_2 \cdot \cos\phi_{20} + d_3 \cdot \cos\phi_{30} = \\ = d_2 \cdot \cos\phi_{20} + d_3 \cdot \cos(\theta + \phi_{20} - \pi) \\ y_M = d_2 \cdot \sin\phi_{20} + d_3 \cdot \sin\phi_{30} = \\ = d_2 \cdot \sin\phi_{20} + d_3 \cdot \sin(\theta + \phi_{20} - \pi) \end{cases} \quad (4)$$

$$\begin{cases} v_M^x \equiv \dot{x}_M = -d_2 \cdot \sin\phi_{20} \cdot \omega_{20} - d_3 \cdot \sin\phi_{30} \cdot \omega_{30} = \\ = -d_2 \cdot \sin\phi_{20} \cdot \omega_{20} - d_3 \cdot \sin\phi_{30} \cdot (\dot\theta + \omega_{20}) \\ v_M^y \equiv \dot{y}_M = d_2 \cdot \cos\phi_{20} \cdot \omega_{20} + d_3 \cdot \cos\phi_{30} \cdot \omega_{30} = \\ = d_2 \cdot \cos\phi_{20} \cdot \omega_{20} + d_3 \cdot \cos\phi_{30} \cdot (\dot\theta + \omega_{20}) \end{cases} \quad (5)$$

$$\begin{cases} a_M^x \equiv \ddot{x}_M = -d_2 \cdot \cos\phi_{20} \cdot \omega_{20}^2 - d_3 \cdot \cos\phi_{30} \cdot \omega_{30}^2 = \\ = -d_2 \cdot \cos\phi_{20} \cdot \omega_{20}^2 - d_3 \cdot \cos\phi_{30} \cdot (\dot\theta + \omega_{20})^2 \\ a_M^y \equiv \ddot{y}_M = -d_2 \cdot \sin\phi_{20} \cdot \omega_{20}^2 - d_3 \cdot \sin\phi_{30} \cdot \omega_{30}^2 = \\ = -d_2 \cdot \sin\phi_{20} \cdot \omega_{20}^2 - d_3 \cdot \sin\phi_{30} \cdot (\dot\theta + \omega_{20})^2 \end{cases} \quad (6)$$

Note: actuator rotation speeds were considered constant (relations 7).

$$\dot\phi_{20} = \omega_{20} = ct; \quad \dot\theta = ct \Rightarrow si \quad \omega_{30} = ct. \quad (7)$$

One takes $\varepsilon_{20} = \ddot\theta = \varepsilon_{30} = 0$.

Relations (3) shall be derived and them and it get the velocities system (8) and the accelerations system (9).

$$\begin{cases} d^2 = x_M^2 + y_M^2 \\ 2 \cdot d \cdot \dot{d} = 2 \cdot x_M \cdot \dot{x}_M + 2 \cdot y_M \cdot \dot{y}_M \\ d \cdot \dot{d} = x_M \cdot \dot{x}_M + y_M \cdot \dot{y}_M \\ \dot{d} = \dfrac{x_M \cdot \dot{x}_M + y_M \cdot \dot{y}_M}{d} \\ d \cdot \cos\phi = x_M; \quad d \cdot \sin\phi = y_M \\ \dot{d} \cdot \cos\phi - d \cdot \sin\phi \cdot \dot\phi = \dot{x}_M \mid \cdot(-\sin\phi) \\ \dot{d} \cdot \sin\phi + d \cdot \cos\phi \cdot \dot\phi = \dot{y}_M \mid \cdot(\cos\phi) \\ d \cdot \dot\phi = \dot{x}_M \cdot (-\sin\phi) + \dot{y}_M \cdot (\cos\phi) \\ \dot\phi = \dfrac{\dot{y}_M \cdot \cos\phi - \dot{x}_M \cdot \sin\phi}{d} \\ \dot{d} = \dfrac{x_M \cdot \dot{x}_M + y_M \cdot \dot{y}_M}{d} \end{cases} \quad (8)$$

$$\begin{cases}
d^2 = x_M^2 + y_M^2 \\
2 \cdot d \cdot \dot{d} = 2 \cdot x_M \cdot \dot{x}_M + 2 \cdot y_M \cdot \dot{y}_M \\
d \cdot \dot{d} = x_M \cdot \dot{x}_M + y_M \cdot \dot{y}_M \\
\dot{d}^2 + d \cdot \ddot{d} = \dot{x}_M^2 + x_M \cdot \ddot{x}_M + \dot{y}_M^2 + y_M \cdot \ddot{y}_M \\
\ddot{d} = \dfrac{\dot{x}_M^2 + x_M \cdot \ddot{x}_M + \dot{y}_M^2 + y_M \cdot \ddot{y}_M - \dot{d}^2}{d} \\
d \cdot \cos\phi = x_M \\
d \cdot \sin\phi = y_M \\
\dot{d} \cdot \cos\phi - d \cdot \sin\phi \cdot \dot{\phi} = \dot{x}_M \mid \cdot(-\sin\phi) \\
\dot{d} \cdot \sin\phi + d \cdot \cos\phi \cdot \dot{\phi} = \dot{y}_M \mid \cdot(\cos\phi) \\
\\ \hline
d \cdot \dot{\phi} = -\dot{x}_M \cdot \sin\phi + \dot{y}_M \cdot \cos\phi \\
\dot{d} \cdot \dot{\phi} + d \cdot \ddot{\phi} = \ddot{y}_M \cdot \cos\phi - \dot{y}_M \cdot \sin\phi \cdot \dot{\phi} - \\
\quad - \ddot{x}_M \cdot \sin\phi - \dot{x}_M \cdot \cos\phi \cdot \dot{\phi} \\
\ddot{\phi} = \dfrac{\ddot{y}_M \cdot \cos\phi - \ddot{x}_M \cdot \sin\phi}{d} + \\
\quad + \dfrac{-\dot{y}_M \cdot \sin\phi \cdot \dot{\phi} - \dot{x}_M \cdot \cos\phi \cdot \dot{\phi} - \dot{d} \cdot \dot{\phi}}{d} \\
\\ \hline
\ddot{d} = \dfrac{\dot{x}_M^2 + x_M \cdot \ddot{x}_M + \dot{y}_M^2 + y_M \cdot \ddot{y}_M - \dot{d}^2}{d}
\end{cases} \qquad (9)$$

The following will be determined positions, speeds and accelerations, according to the positions of scaling point O_3.

Start at the co-ordinates of scaling point O_3 (10).

$$\begin{cases} x_{O_3} = d_2 \cdot \cos\phi_{20} \\ y_{O_3} = d_2 \cdot \sin\phi_{20} \end{cases} \qquad (10)$$

Speeds (Flavio de Melo, 2012) is then determined scaling, and accelerations point O₃, by successive derivation of the system (10), in which shall be replaced in accordance with products diverter d.cos or d.sin with respective positions, xo3 or yo3, which become in this way variables (see relations 11 and 12).

$$\begin{cases} \dot{x}_{O_3} = -d_2 \cdot \sin\phi_{20} \cdot \omega_{20} = -y_{O_3} \cdot \omega_{20} \\ \dot{y}_{O_3} = d_2 \cdot \cos\phi_{20} \cdot \omega_{20} = x_{O_3} \cdot \omega_{20} \end{cases} \quad (11)$$

$$\begin{cases} \ddot{x}_{O_3} = -d_2 \cdot \cos\phi_{20} \cdot \omega_{20}^2 = -x_{O_3} \cdot \omega_{20}^2 \\ \ddot{y}_{O_3} = -d_2 \cdot \sin\phi_{20} \cdot \omega_{20}^2 = -y_{O_3} \cdot \omega_{20}^2 \end{cases} \quad (12)$$

They got so out speeds and accelerations scaling point O₃ depending on the original positions (scale) and the absolute angular speed of the item 2. Angular speed it was considered constant.

Application

Determination of technical gear and acceleration as a function of positions, is especially useful in the study system dynamics, vibration and noise caused by the system (Dong, 2013; Wang, 2013). This technique is often encountered in the study system vibration. Vibrations are known positions of scaling Point O₃ and is to be determined and then easily gear vibration and acceleration point in question as well as the other points on the system as a function of the positions of known scaling point O₃. All through this technique can be calculated local noise levels at various points of the system, as well as overall levels of noise generated by the system to the nearest large enough in comparison with noises obtained by measurements experimental, with appropriate equipment. The study system dynamics can be developed and by this technique.

Absolute velocity of the point O₃ (speed module) is given by the relationship (13).

$$\begin{cases} v_{O_3} = \sqrt{\dot{x}_{O_3}^2 + \dot{y}_{O_3}^2} = \\ = \sqrt{d_2^2 \cdot \omega_{20}^2 \cdot \sin^2\phi_{20} + d_2^2 \cdot \omega_{20}^2 \cdot \cos^2\phi_{20}} = \\ = \sqrt{d_2^2 \cdot \omega_{20}^2} = d_2 \cdot \omega_{20} \end{cases} \quad (13)$$

Absolute acceleration of the point O_3 for angular speed constant, is given by the relationship (14).

$$\begin{cases} a_{O_3} = \sqrt{\ddot{x}_{O_3}^2 + \ddot{y}_{O_3}^2} = \\ = \sqrt{d_2^2 \cdot \omega_{20}^4 \cdot \cos^2 \phi_{20} + d_2^2 \cdot \omega_{20}^4 \cdot \sin^2 \phi_{20}} = \\ = \sqrt{d_2^2 \cdot \omega_{20}^4} = d_2 \cdot \omega_{20}^2 \end{cases} \quad (14)$$

The following parameters (Petrescu, 2009, 2013), cinematics (scalar) of point M, endeffector, will be determined depending on and the parameters of the position points O_3 and M (customer systems 15-17).

$$\begin{cases} x_M = x_{O_3} + d_3 \cdot \cos\phi_{30} \\ y_M = y_{O_3} + d_3 \cdot \sin\phi_{30} \end{cases} \begin{cases} d_3 \cdot \cos\phi_{30} = x_M - x_{O_3} \\ d_3 \cdot \sin\phi_{30} = y_M - y_{O_3} \end{cases} \quad (15)$$

$$\begin{cases} \dot{x}_M = \dot{x}_{O_3} - d_3 \cdot \sin\phi_{30} \cdot \dot{\phi}_{30} = \\ = -y_{O_3} \cdot \omega_{20} + (y_{O_3} - y_M) \cdot (\omega_{20} + \dot{\theta}) = \\ = y_{O_3} \cdot \dot{\theta} - y_M \cdot (\omega_{20} + \dot{\theta}) = \\ = (y_{O_3} - y_M) \cdot \dot{\theta} - y_M \cdot \omega_{20} \\ \dot{y}_M = \dot{y}_{O_3} + d_3 \cdot \cos\phi_{30} \cdot \dot{\phi}_{30} = \\ = x_{O_3} \cdot \omega_{20} + (x_M - x_{O_3}) \cdot (\omega_{20} + \dot{\theta}) = \\ = x_M \cdot (\omega_{20} + \dot{\theta}) - x_{O_3} \cdot \dot{\theta} = \\ = (x_M - x_{O_3}) \cdot \dot{\theta} + x_M \cdot \omega_{20} \\ \dot{y}_{O_3} - \dot{y}_M = -d_3 \cdot \cos\phi_{30} \cdot (\omega_{20} + \dot{\theta}) \\ \dot{x}_M - \dot{x}_{O_3} = -d_3 \cdot \sin\phi_{30} \cdot (\omega_{20} + \dot{\theta}) \end{cases} \quad (16)$$

$$\begin{cases} \ddot{x}_M = (\dot{y}_{O_3} - \dot{y}_M)\cdot\dot{\theta} - \dot{y}_M\cdot\omega_{20} \\ \ddot{y}_M = (\dot{x}_M - \dot{x}_{O_3})\cdot\dot{\theta} + \dot{x}_M\cdot\omega_{20} \\ \dot{y}_{O_3} - \dot{y}_M = (x_{O_3} - x_M)\cdot(\omega_{20} + \dot{\theta}) \\ \dot{x}_M - \dot{x}_{O_3} = (y_{O_3} - y_M)\cdot(\omega_{20} + \dot{\theta}) \\ \ddot{x}_M = (x_{O_3} - x_M)\cdot(\omega_{20} + \dot{\theta})\cdot\dot{\theta} + \\ \quad + (x_{O_3} - x_M)\cdot\dot{\theta}\cdot\omega_{20} - x_M\cdot\omega_{20}^2 \\ \ddot{y}_M = (y_{O_3} - y_M)\cdot(\omega_{20} + \dot{\theta})\cdot\dot{\theta} + \\ \quad + (y_{O_3} - y_M)\cdot\dot{\theta}\cdot\omega_{20} - y_M\cdot\omega_{20}^2 \\ \ddot{x}_M = 2\cdot(x_{O_3} - x_M)\cdot\dot{\theta}\cdot\omega_{20} + \\ \quad + (x_{O_3} - x_M)\cdot\dot{\theta}^2 - x_M\cdot\omega_{20}^2 \\ \ddot{y}_M = 2\cdot(y_{O_3} - y_M)\cdot\dot{\theta}\cdot\omega_{20} + \\ \quad + (y_{O_3} - y_M)\cdot\dot{\theta}^2 - y_M\cdot\omega_{20}^2 \\ \ddot{x}_M = (x_{O_3} - x_M)\cdot(2\cdot\dot{\theta}\cdot\omega_{20} + \dot{\theta}^2) - x_M\cdot\omega_{20}^2 \\ \ddot{y}_M = (y_{O_3} - y_M)\cdot(2\cdot\dot{\theta}\cdot\omega_{20} + \dot{\theta}^2) - y_M\cdot\omega_{20}^2 \\ \ddot{x}_M = (x_{O_3} - x_M)\cdot(\omega_{20} + \dot{\theta})^2 - x_{O_3}\cdot\omega_{20}^2 \\ \ddot{y}_M = (y_{O_3} - y_M)\cdot(\omega_{20} + \dot{\theta})^2 - y_{O_3}\cdot\omega_{20}^2 \end{cases} \quad (17)$$

Conclusions

The method proposed in this chapter has the advantage of greatly simplify day-to-systems design calculations to mechatronics and robotic. Win time, saves work, it's possible to work out a direct synthesis easier serial systems, without it being necessary experimental testing. For teaching method is a total simplification of the design, and thus increasing the understanding of this phenomenon.

References

Dong, H., Giakoumidis, N., Figueroa, N., and Mavridis, N., (2013). Approaching Behaviour Monitor and Vibration Indication in Developing a General Moving Object Alarm System (GMOAS), International Journal of Advanced Robotic Systems, Hanafiah Yussof (Ed.), ISBN: 1729-8806, InTech, DOI: 10.5772/56586. Available from:

http://www.intechopen.com/journals/international_journal_of_advanced_robotic_systems/approaching-behaviour-monitor-and-vibration-indication-in-developing-a-general-moving-object-alarm-s.

Flavio de Melo, L., Reis Alves, S.F., Rosário, J.M., Mobile Robot Navigation Modelling, Control and Applications, in International Review on Modelling and Simulations, April 2012, Vol. 5, N. 2B, pp. 1059-1068. Available from:

http://www.praiseworthyprize.com/IREMOS-latest/IREMOS_vol_5_n_2.html.

Lee, B.J., (2013). Geometrical Derivation of Differential Kinematics to Calibrate Model Parameters of Flexible Manipulator, International Journal of Advanced Robotic Systems, Jaime Gallardo-Alvarado, Ramon Rodríguez-Castro (Ed.), ISBN: 1729-8806, InTech, DOI: 10.5772/55592. Available from: http://www.intechopen.com/journals/international_journal_of_advanced_robotic_systems/geometrical-derivation-of-differential-kinematics-to-calibrate-model-parameters-of-flexible-manipula.

Garcia, E.; Jimenez, M.A.; De Santos, P.G.; Armada, M., "The evolution of robotics research," Robotics & Automation Magazine, IEEE , vol.14, no.1, pp.90,103, March 2007. Available from:

http://ieeexplore.ieee.org/stamp/stamp.jsp?tp=&arnumber=4141037&isnumber=4141014.

He, B., Wang, Z., Li, Q., Xie, H., and Shen, R., (2013). An Analytic Method for the Kinematics and Dynamics of a Multiple-Backbone Continuum Robot, IJARS, Patricia Melin (Ed.), ISBN: 1729-8806, InTech, DOI: 10.5772/54051. Available from:

http://www.intechopen.com/journals/international_journal_of_advanced_robotic_systems/an-analytic-method-for-the-kinematics-and-dynamics-of-a-multiple-backbone-continuum-robot.

Liu, H., Zhou, W., Lai, X., and Zhu, S., (2013). An Efficient Inverse Kinematic Algorithm for a PUMA560-Structured Robot Manipulator, IJARS, Jaime Gallardo-Alvarado, Ramon Rodríguez-Castro (Ed.), ISBN: 1729-8806, InTech, DOI: 10.5772/56403. Available from:

http://www.intechopen.com/journals/international_journal_of_advanced_robotic_systems/an-efficient-inverse-kinematic-algorithm-for-a-puma560-structured-robot-manipulator.

Garcia-Murillo, M., Gallardo-Alvarado, J., and Castillo-Castaneda, E., (2013). Finding the Generalized Forces of a Series-Parallel Manipulator, IJARS, Jaime Gallardo-Alvarado, Ramon Rodrıguez-Castro (Ed.), ISBN: 1729-8806, InTech, DOI: 10.5772/53824. Available from:

http://www.intechopen.com/journals/international_journal_of_advanced_robotic_systems/finding-the-generalized-forces-of-a-series-parallel-manipulator.

Padula, F., and Perdereau, V., (2013). An On-Line Path Planner for Industrial Manipulators, International Journal of Advanced Robotic Systems, Antonio Visioli (Ed.), ISBN: 1729-8806, InTech, DOI: 10.5772/55063. Available from:

http://www.intechopen.com/journals/international_journal_of_advanced_robotic_systems/an-on-line-path-planner-for-industrial-manipulators.

Perumaal, S., and Jawahar, N., (2013). Automated Trajectory Planner of Industrial Robot for Pick-and-Place Task, IJARS, Antonio Visioli (Ed.), ISBN: 1729-8806, InTech, DOI: 10.5772/53940. Available from:

http://www.intechopen.com/journals/international_journal_of_advanced_robotic_systems/automated-trajectory-planner-of-industrial-robot-for-pick-and-place-task.

Petrescu, F.I., Petrescu, R.V., Cinematics of the 3R Dyad, in journal Engevista, vol. 15, No. 2, (2013), August 2013, ISSN 1415-7314, pp. 118-124. Available from:

http://www.uff.br/engevista/seer/index.php/engevista/article/view/376.

Petrescu, F.I., Petrescu, R.V., Kinematics of the Planar Quadrilateral Mechanism, in journal Engevista, vol. 14, No. 3, (2012), December 2012, ISSN 1415-7314, pp. 345-348. Available from:

http://www.uff.br/engevista/seer/index.php/engevista/article/view/377.

Petrescu, F.I., Petrescu, R.V., Mecatronica – Sisteme Seriale si Paralele, Create Space publisher, USA, March 2012, ISBN 978-1-4750-6613-5, 128 pages, Romanian edition.

Petrescu, F.I, Petrescu, R.V, Mechanical Systems, Serial and Parallel – Course (in romanian), LULU Publisher, London, UK, February 2011, 124 pages, ISBN 978-1-4466-0039-9, Romanian edition.

Petrescu, F.I., Grecu, B., Comanescu, A., Petrescu, R.V., Some Mechanical Design Elements. In the 3rd International Conference on Computational Mechanics and Virtual Engineering, COMEC 2009, Braşov, October 2009, ISBN 978-973-598-572-1, Edit. UTB, pp. 520-525.

Reddy, P., Shihabudheen K.V., Jacob, J., Precise Non Linear Modeling of Flexible Link Flexible Joint Manipulator, in International Review on Modelling and Simulations, June 2012, Vol. 5, N. 3B, pp. 1368-1374. Available from:

http://www.praiseworthyprize.com/IREMOS-latest/IREMOS_vol_5_n_3.html.

Tang, X., Sun, D., and Shao, Z., (2013). The Structure and Dimensional Design of a Reconfigurable PKM, IJARS, Sumeet S Aphale (Ed.), ISBN: 1729-8806, InTech, DOI: 10.5772/54696. Available from:

http://www.intechopen.com/journals/international_journal_of_advanced_robotic_systems/the-structure-and-dimensional-design-of-a-reconfigurable-pkm.

Tong, G., Gu, J., and Xie, W., (2013). Virtual Entity-Based Rapid Prototype for Design and Simulation of Humanoid Robots, International Journal of Advanced Robotic Systems, Guangming Xie (Ed.), ISBN: 1729-8806, InTech, DOI: 10.5772/55936. Available from:

http://www.intechopen.com/journals/international_journal_of_advanced_robotic_systems/virtual-entity-based-rapid-prototype-for-design-and-simulation-of-humanoid-robots.

Wang, K., Luo, M., Mei, T., Zhao, J., and Cao, Y., (2013). Dynamics Analysis of a Three-DOF Planar Serial-Parallel Mechanism for Active Dynamic Balancing with Respect to a Given Trajectory, International Journal of Advanced Robotic Systems, Sumeet S Aphale (Ed.), ISBN: 1729-8806, InTech, DOI: 10.5772/54201. Available from:

http://www.intechopen.com/journals/international_journal_of_advanced_robotic_systems/dynamics-analysis-of-a-three-dof-planar-serial-parallel-mechanism-for-active-dynamic-balancing-with-.

Wen, S., Zhu, J., Li, X., Rad, A., Chen, X., (2012). End-Point Contact Force Control with Quantitative Feedback Theory for Mobile Robots, IJARS, Houxiang Zhang, Shengyong Chen (Ed.), ISBN: 1729-8806, InTech, DOI: 10.5772/53742. Available from:

http://www.intechopen.com/journals/international_journal_of_advanced_robotic_systems/end-point-contact-force-control-with-quantitative-feedback-theory-for-mobile-robots.

Chapter 04_Inverse Kinematics of Moving (Serial) Mechanical Systems by a Geometric Method

Introduction

Moving mechanical structures are used increasingly in almost all vital sectors of humanity [17]. The robots are able to process integrated circuits sizes micro and nano, on which the man they can be seen even with electron microscopy [3]. Dyeing parts in toxic environments, working in chemical and radioactive environments, or at depths and pressures at the bottom of huge oceans, or even cosmic space conquest and visiting exo-planets, are now possible, and were turned into from the dream in reality, because mechanical platforms sequential gearbox [1].

The man will be able to carry out its mission supreme, conqueror of new galaxies, because mechanical systems sequential gearbox [9].

Robots were developed and diversified, different aspects, but today, they start to be directed on two major categories: systems serial and parallel systems [8].

Parallel systems are more solid, but more difficult to designed and handled, which serial systems were those which have developed the most.

Serial systems [15] and they have different constructive diagrams, but over the last 30 years have been channelled on anthropomorphic structures [1-17].

These structures are made up of simple components and couplers for rotation.

Their great advantage is fast movements, good dynamics, a high accuracy, a construction of simple modules STAS, economy of materials, low cost, and high reliability.

One disadvantage of less accurate has been removed because of stepper motors.

Compared with parallel systems (more solid but more cumbersome) serial systems may pose, and the disadvantage of stability something lower.

This disadvantage begins to be exceeded today in smart mode, through the construction of serial systems made up of elements doubled (in parallel).

This last invention, will lead to the strengthening of serial systems, and to their consolidation like the indisputable leader in diversity of mechatronics and robotic systems.

This work starts from a main idea, to study these systems on a single model, 3R, which has finally main movements lying on a single plane model, 2R.

The inverse kinematics (plan 2-3)

In figure 1 can be monitored kinematic chain schematic plan 2-3 opened.

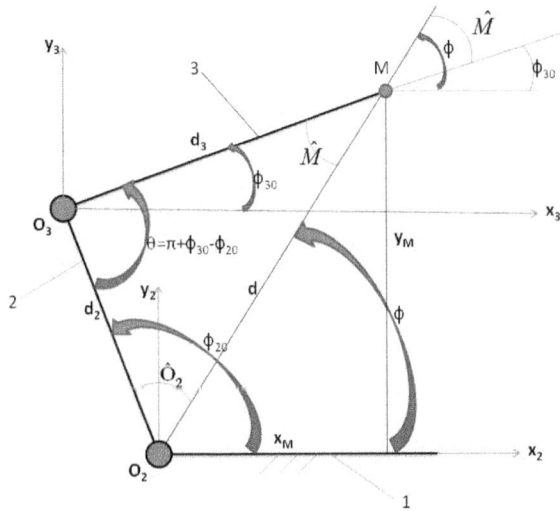

Figure 1. *Cinematic schematic kinematic chain plan 2-3 linked to item 1 considered fixed*

In the direct kinematics [18] are known the parameters cinematic Φ_{20} and Φ_{30} and must be determined by analytical calculation parameters x_M and y_M, which represents the co-ordinates of scaling point M (endeffector M). Designing vectors $d_2 + d_3$ on cartesian axis system considered fixed, xOy, identical to $x_2O_2y_2$. Obtain the equation scaling (2).

In inverse kinematics are known the coordinates of point M, x_M and y_M, and the dimensions d_2 and d_3, and must be determined by analytical calculation the cinematic parameters φ_{20} and φ_{30}.

The geometric method (which will be presented) determines first the supplementaires parameters x_{O_3}, y_{O_3}.

To be known and parameter d variable-length, which is calculated by the relation 2 [2].

47

$$\begin{cases} d^2 = x_M^2 + y_M^2 \\ d = \sqrt{d^2} = \sqrt{x_M^2 + y_M^2} \end{cases} \quad (2)$$

Note the parameters x_{O_3}, y_{O_3}, with x, y (see the relation 3).

$$\begin{cases} x_{O_3} = x \\ y_{O_3} = y \end{cases} \quad (3)$$

It is possible to write the equations system 4.

$$\begin{cases} (x - x_{O_2})^2 + (y - y_{O_2})^2 = d_2^2 \\ (x - x_M)^2 + (y - y_M)^2 = d_3^2 \end{cases} \quad (4)$$

Analytical equations 4 shall be written as either two circles of the ray d2 and d3, either by the use of two right triangles (using the Pythagorean relationship). Look at parameters in fixed cartesian system x₂O₂y₂. And by reversing the equations, the system 4 takes the form 5.

$$\begin{cases} (x - x_M)^2 + (y - y_M)^2 = d_3^2 \\ x^2 + y^2 = d_2^2 \end{cases} \quad (5)$$

The squared first equation and 5 system acquires the form 6 (It has been entered, and the relationship 2).

$$\begin{cases} x^2 + y^2 + x_M^2 + y_M^2 - d_3^2 - 2x_M x - 2y_M y = 0 \\ x^2 + y^2 = d_2^2 \\ x_M^2 + y_M^2 = d^2 \end{cases} \quad (6)$$

Using two and three expressions, first relationship of the system 6 takes shape 7.

$$2x_M x = d_2^2 + d^2 - d_3^2 - 2y_M y \quad (7)$$

We can now let it explicit the x (relationship 8).

$$x = \frac{d^2 + d_2^2 - d_3^2 - 2y_M y}{2x_M} \quad (8)$$

We square the expression 8 and we obtain the value of x square (relationship 9).

$$x^2 = \frac{[(d^2+d_2^2-d_3^2)-2y_My]^2}{4x_M^2} =$$
$$= \frac{(d^2+d_2^2-d_3^2)^2 + 4y_M^2 y^2 - 4y_M(d^2+d_2^2-d_3^2)y}{4x_M^2} \qquad (9)$$

Write now 5-6 in x square (see system 10).

$$\begin{cases} x^2 = \dfrac{(d^2+d_2^2-d_3^2)^2 + 4y_M^2 y^2 - 4y_M(d^2+d_2^2-d_3^2)y}{4x_M^2} \\ x^2 = d_2^2 - y^2 \end{cases} \qquad (10)$$

Result the equation 11.

$$\frac{(d^2+d_2^2-d_3^2)^2 + 4y_M^2 y^2 - 4y_M(d^2+d_2^2-d_3^2)y}{4x_M^2} = d_2^2 - y^2 \qquad (11)$$

Which take the form 12.

$$(d^2+d_2^2-d_3^2)^2 + 4y_M^2 y^2 - 4y_M(d^2+d_2^2-d_3^2)y = \\ = 4x_M^2 d_2^2 - 4x_M^2 y^2 \qquad (12)$$

The equation 12 is placed under simplified form 13.

$$4d^2 y^2 - 4y_M(d^2+d_2^2-d_3^2)y - \\ - 4x_M^2 d_2^2 + (d^2+d_2^2-d_3^2)^2 = 0 \qquad (13)$$

Now we can determine directly the two solutions (see the system 14).

$$\begin{cases} y_{O_3} \equiv y_{1,2} = \dfrac{y_M\left(d^2 + d_2^2 - d_3^2\right)}{2d^2} \pm \dfrac{\sqrt{\Delta}}{2d^2} \\ x_{O_3} \equiv x_{1,2} = \dfrac{\left(d^2 + d_2^2 - d_3^2\right) - 2y_M y_{1,2}}{2x_M} \\ \Delta = 4d^2 d_2^2 x_M^2 + \left(d^2 + d_2^2 - d_3^2\right)^2 \cdot \left(y_M^2 - d^2\right) \end{cases} \quad (14)$$

The two solutions correspond to the two possible positions (see the Figure 2). Plus sign corresponds to the position Nordic, O_3. Minus sign corresponds to the position Southern O_3'.

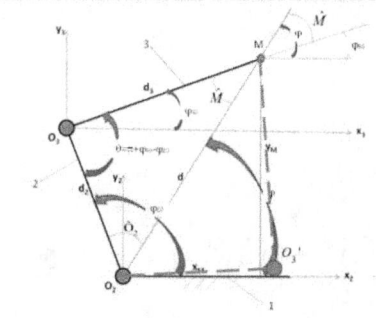

Figure 2. *Cinematic schematic kinematic chain plan 2-3 linked to item 1 considered fixed; There are two possible physical positions*

Once it is determined the point O_3 can be calculated immediately the two angles of positions, φ_{20} and φ_{30} (systems 15 and 16).

$$\begin{cases} x_{O_3} = x_{O_2} + d_2 \cdot \cos\varphi_{20} \Rightarrow \cos\varphi_{20} = \dfrac{x_{O_3}}{d_2} \\ y_{O_3} = y_{O_2} + d_2 \cdot \sin\varphi_{20} \Rightarrow \sin\varphi_{20} = \dfrac{y_{O_3}}{d_2} \\ \varphi_{20} = \text{sign}(\sin\varphi_{20}) \cdot \arccos(\cos\varphi_{20}) \\ \varphi_{20} = \text{sign}\left(\dfrac{y_{O_3}}{d_2}\right) \cdot \arccos\left(\dfrac{x_{O_3}}{d_2}\right) \end{cases} \quad (15)$$

$$\begin{cases} x_M = x_{O_3} + d_3 \cdot \cos\varphi_{30} \Rightarrow \cos\varphi_{30} = \dfrac{x_M - x_{O_3}}{d_3} \\ y_M = y_{O_3} + d_3 \cdot \sin\varphi_{30} \Rightarrow \sin\varphi_{30} = \dfrac{y_M - y_{O_3}}{d_3} \\ \varphi_{30} = sign(\sin\varphi_{30}) \cdot \arccos(\cos\varphi_{30}) \\ \varphi_{30} = sign\left(\dfrac{y_M - y_{O_3}}{d_3}\right) \cdot \arccos\left(\dfrac{x_M - x_{O_3}}{d_3}\right) \end{cases} \quad (16)$$

The system (17) [10, 14] shall be derived as a function of time, resulting in the system (18), which derived with time generates in turn the accelerations (19) [1, 18].

$$\begin{cases} x_M = d_2 \cdot \cos\phi_{20} + d_3 \cdot \cos\phi_{30} = d_2 \cdot \cos\phi_{20} + d_3 \cdot \cos(\theta + \phi_{20} - \pi) \\ y_M = d_2 \cdot \sin\phi_{20} + d_3 \cdot \sin\phi_{30} = d_2 \cdot \sin\phi_{20} + d_3 \cdot \sin(\theta + \phi_{20} - \pi) \end{cases} \quad (17)$$

$$\begin{cases} v_M^x \equiv \dot{x}_M = -d_2 \cdot \sin\phi_{20} \cdot \omega_{20} - d_3 \cdot \sin\phi_{30} \cdot \omega_{30} = \\ \quad = -d_2 \cdot \sin\phi_{20} \cdot \omega_{20} - d_3 \cdot \sin\phi_{30} \cdot (\dot{\theta} + \omega_{20}) \\ v_M^y \equiv \dot{y}_M = d_2 \cdot \cos\phi_{20} \cdot \omega_{20} + d_3 \cdot \cos\phi_{30} \cdot \omega_{30} = \\ \quad = d_2 \cdot \cos\phi_{20} \cdot \omega_{20} + d_3 \cdot \cos\phi_{30} \cdot (\dot{\theta} + \omega_{20}) \end{cases} \quad (18)$$

$$\begin{cases} a_M^x \equiv \ddot{x}_M = -d_2 \cdot \cos\phi_{20} \cdot \omega_{20}^2 - d_3 \cdot \cos\phi_{30} \cdot \omega_{30}^2 = \\ \quad = -d_2 \cdot \cos\phi_{20} \cdot \omega_{20}^2 - d_3 \cdot \cos\phi_{30} \cdot (\dot{\theta} + \omega_{20})^2 \\ a_M^y \equiv \ddot{y}_M = -d_2 \cdot \sin\phi_{20} \cdot \omega_{20}^2 - d_3 \cdot \sin\phi_{30} \cdot \omega_{30}^2 = \\ \quad = -d_2 \cdot \sin\phi_{20} \cdot \omega_{20}^2 - d_3 \cdot \sin\phi_{30} \cdot (\dot{\theta} + \omega_{20})^2 \end{cases} \quad (19)$$

Note: actuator rotation speeds were considered constant (relations 20).

$$\dot{\phi}_{20} = \omega_{20} = ct; \quad \dot{\theta} = ct \Rightarrow si \quad \omega_{30} = ct. \quad (20)$$

One takes $\varepsilon_{20} = \ddot{\theta} = \varepsilon_{30} = 0.$

Relations (2) shall be derived and it get the velocities system (21) and the accelerations system (22).

$$\begin{cases} d^2 = x_M^2 + y_M^2; \quad 2 \cdot d \cdot \dot{d} = 2 \cdot x_M \cdot \dot{x}_M + 2 \cdot y_M \cdot \dot{y}_M \\ d \cdot \dot{d} = x_M \cdot \dot{x}_M + y_M \cdot \dot{y}_M \Rightarrow \dot{d} = \dfrac{x_M \cdot \dot{x}_M + y_M \cdot \dot{y}_M}{d} \\ d \cdot \cos\phi = x_M; \quad d \cdot \sin\phi = y_M \\ \dot{d} \cdot \cos\phi - d \cdot \sin\phi \cdot \dot{\phi} = \dot{x}_M \mid \cdot(-\sin\phi) \\ \dot{d} \cdot \sin\phi + d \cdot \cos\phi \cdot \dot{\phi} = \dot{y}_M \mid \cdot(\cos\phi) \\ d \cdot \dot{\phi} = \dot{x}_M \cdot (-\sin\phi) + \dot{y}_M \cdot (\cos\phi) \\ \dot{\phi} = \dfrac{\dot{y}_M \cdot \cos\phi - \dot{x}_M \cdot \sin\phi}{d}; \quad \dot{d} = \dfrac{x_M \cdot \dot{x}_M + y_M \cdot \dot{y}_M}{d} \end{cases} \quad (21)$$

$$\begin{cases} d^2 = x_M^2 + y_M^2 \\ 2 \cdot d \cdot \dot{d} = 2 \cdot x_M \cdot \dot{x}_M + 2 \cdot y_M \cdot \dot{y}_M \\ d \cdot \dot{d} = x_M \cdot \dot{x}_M + y_M \cdot \dot{y}_M \\ \dot{d}^2 + d \cdot \ddot{d} = \dot{x}_M^2 + x_M \cdot \ddot{x}_M + \dot{y}_M^2 + y_M \cdot \ddot{y}_M \\ \ddot{d} = \dfrac{\dot{x}_M^2 + x_M \cdot \ddot{x}_M + \dot{y}_M^2 + y_M \cdot \ddot{y}_M - \dot{d}^2}{d} \\ d \cdot \cos\phi = x_M; \quad d \cdot \sin\phi = y_M \\ \dot{d} \cdot \cos\phi - d \cdot \sin\phi \cdot \dot{\phi} = \dot{x}_M \mid \cdot(-\sin\phi) \\ \dot{d} \cdot \sin\phi + d \cdot \cos\phi \cdot \dot{\phi} = \dot{y}_M \mid \cdot(\cos\phi) \\ d \cdot \dot{\phi} = -\dot{x}_M \cdot \sin\phi + \dot{y}_M \cdot \cos\phi \\ \dot{d} \cdot \dot{\phi} + d \cdot \ddot{\phi} = \ddot{y}_M \cdot \cos\phi - \dot{y}_M \cdot \sin\phi \cdot \dot{\phi} - \\ \quad - \ddot{x}_M \cdot \sin\phi - \dot{x}_M \cdot \cos\phi \cdot \dot{\phi} \\ \ddot{\phi} = \dfrac{\ddot{y}_M \cdot \cos\phi - \ddot{x}_M \cdot \sin\phi}{d} + \\ \quad + \dfrac{-\dot{y}_M \cdot \sin\phi \cdot \dot{\phi} - \dot{x}_M \cdot \cos\phi \cdot \dot{\phi} - \dot{d} \cdot \dot{\phi}}{d} \\ \ddot{d} = \dfrac{\dot{x}_M^2 + x_M \cdot \ddot{x}_M + \dot{y}_M^2 + y_M \cdot \ddot{y}_M - \dot{d}^2}{d} \end{cases} \quad (22)$$

Conclusions

The method proposed in this work has the advantage of greatly simplify day-to-systems design calculations to mechatronics and robotic. Win time, saves work, it's possible to work out a direct synthesis easier serial systems, without it being necessary experimental testing. For teaching method is a total simplification of the design, and thus increasing the understanding of this phenomenon.

References

[1] Dong, H., Giakoumidis, N., Figueroa, N., and Mavridis, N., (2013). Approaching Behaviour Monitor and Vibration Indication in Developing a General Moving Object Alarm System (GMOAS), International Journal of Advanced Robotic Systems, Hanafiah Yussof (Ed.), ISBN: 1729-8806, InTech, DOI: 10.5772/56586. Available from:

http://www.intechopen.com/journals/international_journal_of_advanced_ro botic_systems/approaching-behaviour-monitor-and-vibration-indication-in-developing-a-general-moving-object-alarm-s.

[2] Flavio de Melo, L., Reis Alves, S.F., Rosário, J.M., Mobile Robot Navigation Modelling, Control and Applications, in International Review on Modelling and Simulations, April 2012, Vol. 5, N. 2B, pp. 1059-1068. Available from:

http://www.praiseworthyprize.com/IREMOS-latest/IREMOS_vol_5_n_2.html.

[3] Lee, B.J., (2013). Geometrical Derivation of Differential Kinematics to Calibrate Model Parameters of Flexible Manipulator, International Journal of Advanced Robotic Systems, Jaime Gallardo-Alvarado, Ramon Rodriguez-Castro (Ed.), ISBN: 1729-8806, InTech, DOI: 10.5772/55592. Available from:

http://www.intechopen.com/journals/international_journal_of_advanced_ro botic_systems/geometrical-derivation-of-differential-kinematics-to-calibrate-model-parameters-of-flexible-manipula.

[4] Garcia, E.; Jimenez, M.A.; De Santos, P.G.; Armada, M., "The evolution of robotics research," Robotics & Automation Magazine, IEEE , vol.14, no.1, pp.90,103, March 2007. Available from:

http://ieeexplore.ieee.org/stamp/stamp.jsp?tp=&arnumber=4141037&isnu mber=4141014.

[5] He, B., Wang, Z., Li, Q., Xie, H., and Shen, R., (2013). An Analytic Method for the Kinematics and Dynamics of a Multiple-Backbone Continuum Robot, IJARS, Patricia Melin (Ed.), ISBN: 1729-8806, InTech, DOI: 10.5772/54051. Available from:

http://www.intechopen.com/journals/international_journal_of_advanced_robotic_systems/an-analytic-method-for-the-kinematics-and-dynamics-of-a-multiple-backbone-continuum-robot.

[6] Liu, H., Zhou, W., Lai, X., and Zhu, S., (2013). An Efficient Inverse Kinematic Algorithm for a PUMA560-Structured Robot Manipulator, IJARS, Jaime Gallardo-Alvarado, Ramon Rodrıguez-Castro (Ed.), ISBN: 1729-8806, InTech, DOI: 10.5772/56403. Available from:

http://www.intechopen.com/journals/international_journal_of_advanced_robotic_systems/an-efficient-inverse-kinematic-algorithm-for-a-puma560-structured-robot-manipulator.

[7] Garcia-Murillo, M., Gallardo-Alvarado, J., and Castillo-Castaneda, E., (2013). Finding the Generalized Forces of a Series-Parallel Manipulator, IJARS, Jaime Gallardo-Alvarado, Ramon Rodrıguez-Castro (Ed.), ISBN: 1729-8806, InTech, DOI: 10.5772/53824. Available from:

http://www.intechopen.com/journals/international_journal_of_advanced_robotic_systems/finding-the-generalized-forces-of-a-series-parallel-manipulator.

[8] Padula, F., and Perdereau, V., (2013). An On-Line Path Planner for Industrial Manipulators, International Journal of Advanced Robotic Systems, Antonio Visioli (Ed.), ISBN: 1729-8806, InTech, DOI: 10.5772/55063. Available from:

http://www.intechopen.com/journals/international_journal_of_advanced_robotic_systems/an-on-line-path-planner-for-industrial-manipulators.

[9] Perumaal, S., and Jawahar, N., (2013). Automated Trajectory Planner of Industrial Robot for Pick-and-Place Task, IJARS, Antonio Visioli (Ed.), ISBN: 1729-8806, InTech, DOI: 10.5772/53940. Available from:

http://www.intechopen.com/journals/international_journal_of_advanced_robotic_systems/automated-trajectory-planner-of-industrial-robot-for-pick-and-place-task.

[10] Petrescu, F.I., Petrescu, R.V., Cinematics of the 3R Dyad, in journal Engevista, vol. 15, No. 2, (2013), August 2013, ISSN 1415-7314, pp. 118-124. Available from:

http://www.uff.br/engevista/seer/index.php/engevista/article/view/376.

[11] Petrescu, F.I., Petrescu, R.V., Kinematics of the Planar Quadrilateral Mechanism, in journal Engevista, vol. 14, No. 3, (2012), December 2012, ISSN 1415-7314, pp. 345-348. Available from:

http://www.uff.br/engevista/seer/index.php/engevista/article/view/377.

[12] Petrescu, F.I., Petrescu, R.V., Mecatronica – Sisteme Seriale si Paralele, Create Space publisher, USA, March 2012, ISBN 978-1-4750-6613-5, 128 pages, Romanian edition.

[13] Petrescu, F.I, Petrescu, R.V, Mechanical Systems, Serial and Parallel – Course (in romanian), LULU Publisher, London, UK, February 2011, 124 pages, ISBN 978-1-4466-0039-9, Romanian edition.

[14] Petrescu, F.I., Grecu, B., Comanescu, A., Petrescu, R.V., Some Mechanical Design Elements. In the 3rd International Conference on Computational Mechanics and Virtual Engineering, COMEC 2009, Braşov, October 2009, ISBN 978-973-598-572-1, Edit. UTB, pp. 520-525.

[15] Reddy, P., Shihabudheen K.V., Jacob, J., Precise Non Linear Modeling of Flexible Link Flexible Joint Manipulator, in International Review on Modelling and Simulations, June 2012, Vol. 5, N. 3B, pp. 1368-1374. Available from:

http://www.praiseworthyprize.com/IREMOS-latest/IREMOS_vol_5_n_3.html.

[16] Tang, X., Sun, D., and Shao, Z., (2013). The Structure and Dimensional Design of a Reconfigurable PKM, IJARS, Sumeet S Aphale (Ed.), ISBN: 1729-8806, InTech, DOI: 10.5772/54696. Available from:

http://www.intechopen.com/journals/international_journal_of_advanced_ro botic_systems/the-structure-and-dimensional-design-of-a-reconfigurable-pkm.

[17] Tong, G., Gu, J., and Xie, W., (2013). Virtual Entity-Based Rapid Prototype for Design and Simulation of Humanoid Robots, International Journal of Advanced Robotic Systems, Guangming Xie (Ed.), ISBN: 1729-8806, InTech, DOI: 10.5772/55936. Available from:

http://www.intechopen.com/journals/international_journal_of_advanced_ro botic_systems/virtual-entity-based-rapid-prototype-for-design-and-simulation-of-humanoid-robots.

[18] Wang, K., Luo, M., Mei, T., Zhao, J., and Cao, Y., (2013). Dynamics Analysis of a Three-DOF Planar Serial-Parallel Mechanism for Active Dynamic Balancing with Respect to a Given Trajectory, International Journal of Advanced Robotic Systems, Sumeet S Aphale (Ed.), ISBN: 1729-8806, InTech, DOI: 10.5772/54201. Available from:

http://www.intechopen.com/journals/international_journal_of_advanced_ro botic_systems/dynamics-analysis-of-a-three-dof-planar-serial-parallel-mechanism-for-active-dynamic-balancing-with-.

Chapter 05_ Inverse Kinematics to the Anthropomorphic Robots, by the Trigonometric Method

Introduction

Moving mechanical structures are used increasingly in almost all vital sectors of humanity [18]. The robots are able to process integrated circuits [1] sizes micro and nano, on which the man they can be seen even with electron microscopy [4]. Dyeing parts in toxic environments, working in chemical and radioactive environments, or at depths and pressures at the bottom of huge oceans, or even cosmic space conquest and visiting exo-planets, are now possible, and were turned into from the dream in reality, because mechanical platforms sequential gearbox [2].

The man will be able to carry out its mission supreme, conquer-or of new galaxies [3], because mechanical systems sequential gear-box [10].
Robots were developed and diversified, different aspects, but to-day, they start to be directed on two major categories: systems serial and parallel systems [9].

Parallel systems are more solid, but more difficult to designed and handled, which serial systems were those which have devel-oped the most.
Serial systems and they have different constructive diagrams, but over the last 30 years have been channelled on anthropomorphic structures [16].

These structures are made up of simple components and cou-plers for rotation.

Their great advantage is fast movements, good dynamics, a high accuracy, a construction of simple modules STAS, economy of materials, low cost, and high reliability.

One disadvantage of less accurate has been removed because of stepper motors.

Compared with parallel systems (more solid but more cumber-some) serial systems may pose, and the disadvantage of stability something lower.

This disadvantage begins to be exceeded today in smart mode, through the construction of serial systems made up of elements doubled (in parallel).

This last invention, will lead to the strengthening of serial sys-tems, and to their consolidation like the indisputable leader in diversity of mechatronics and robotic systems.

This work starts from a main idea, to study these systems on a single model, 3R, which has finally main movements lying on a single plane model, 2R.

The most commonly used serial structures over the last 20 or 30 years are those of type 3R, 4R, 5R, 6R, having as constituents essential basic kinematic chain 3R, robot antropomorf (RRR), where main rotation around a vertical axis, causes the construction [12-13].

There are then a basic kinematic chain which has two revolutions 'bokeh' (two actuators, i.e. two motors) who work permanently in one plane, and immediately after main support which supports and rotates vertically complete assembly [14].

This basic structure [17], 3R, a meet me at all robots serial manufactured on the principle of rotations. Vertical Bracket is was the same, but the drive train as follows, with the two turns situated in a plane can be positioned vertically (most often; the robots anthropomorphic), or horizontally.

It can thus passes from the study spatial movement, which is more difficult, to the study motion plane, basic movement, for all the robots and fillers serial movements of rotation.

Moving flat, horizontal or vertical, shall be undertaken far more easily than the spatial integration with the convenience simple in the space of which it is part [5].

The basic structure [11-15] existing in a few serial platforms of rotation, these being the most generalized (more widespread) at the present time. On this basic model (3R) have developed further robots 6R [6] today (rely only on revolutions using actuator that only electric motors, compact); they have a hardness greater penetration while maintaining the flexibility and models 3R, 4R and 5R.

Almost all major companies come today with models 6R (which they improve continuously).

Why they have imposed today these models of robots (after tens of years of diversity was the word of order); may and of the need for standardisation, or to find a common solution, after a huge portion (however are not yet the only robots use of category serial, but they also have the widest spread) [11-15].

The six turns (full elimination of translations, who bring many disadvantages due to coupler T itself) are operating easier, faster, with higher yield, more reliable, more compact and more secure; basic revolutions, remain all first three, the other three turns (additional) having the role of position may well end device, the endeffector. Results and that the baseline study (required) it is still not for a 3R [11-15].

This can be seen in the latest models of the various firms producing robots (Kawasaki, Romat, Fanuc, Motoman, Kuka, etc). And the structures used inside cells sequential gearbox are constructed generally in a similar way.

Starting from the basic 3R platform may be studied by addition any other scheme, n-R modern [11-15].

The platform (system), has three degree of mobility, which can be realized by three actuators (electric motors). First electric motor drives your whole system in a rotating around a vertical spindle O0z0. Engine (actuator) number 1, is mounted on the fixed (frame, 0) and causes mobile element 1 in a rotating around a vertical axis. The mobile element 1, then all the other elements (components) of the system.

Follows a kinematic chain plane (vertical), composed of two moving components and two couplers 'bokeh' engines. It's about the kinematic elements mobile 2 and 3, the assembly 2-3 being moved by the actuator of the second mounted to engage A (O_2), fixed on item 1. Therefore the second electric motor attached to the component 1 will result in the item 2 of rotating relative to the item 1, but he will move automatically entire drive train 2-3 [11-15].

Last actuator (electric motor) attached to the item (2), and in B (O_3), will rotate item 3 (relative in relation to 2). Rotation φ_{10} carried out by the first actuator, is and relative (between items 1 and 0) and absolute (between elements 1 and 0). Rotation φ_{20} carried out by the second actuator, is and relative (between items 2 and 1) and absolute (between items 2 and 0), due to the arrangement. Rotation $\theta=\varphi_{32}$ carried out by the third actuator, is only relative (between items 3 and 2); the corresponding absolute (between items 3 and 0) as a function of $\theta=\varphi_{32}$ and φ_{20}.

The drive train 2-3 (consisting of kinematic elements 2 and 3) is a kinematic chain plan, which fall within a single plane or in one or more of the other plane parallel to each other. It is a special kinematic system, which will be examined separately. It shall be considered as item 1 of which is caught the drive train 2-3 as being fixed, couplers kinematic engines A(O_2) and B(O_3) becoming first fixed coupler, and the coupler to make two mobile, which are both couplers kinematic C5, of rotation [11-15].

Determination of the extent of mobility of kinematic chain plan 2-3, apply structural formula given by the relationship ($M_3 = 3 \cdot m - 2 \cdot C_5 = 3 \cdot 2 - 2 \cdot 2 = 6 - 4 = 2$), where m represents the number of elements of the kinematic chain mobile plan, in our case m=2 (as we are talking about the two kinematic elements element noted with 2 and 3), and C_5 represents the number of couplers 'bokeh' fifth-class, in the case in point C_5=2 (in the case of couplers A and B or O_2 and O_3).

The drive train 2-3 having degree of mobility 2, must be actuated by two motors [7].

It is preferred that the two actuators to be two electric motors, a direct current, or alternately. Action can be done but with different engines. Hydraulic motors, pneumatic, sonic, etc.

Structural schematic kinematic chain plan 2-3 resembles with cinematic scheme.

The conductor 2 is linked to the element considered fixed coupler 1 by the term O2 engines, and drive element 3 is connected to the element mobile 2 by engage engines O3. This results in a kinematic chain open with two degree of mobility, which can be realized by the two actuators, i.e. the two electric motors, mounted on 'bokeh' couplers engines A and B or O2 or O3 [8]. In the direct kinematics [19] are known the parameters cinematic Φ20 and Φ30 and must be determined by analytical calculation parameters x_M and y_M, which represents the co-ordinates of scaling point M (endeffector M). Designing vectors d2 + d3 on cartesian axis system considered fixed, xOy, identical to $x_2O_2y_2$. Obtain the equation scaling (2) [20].

Anthropomorphic robots [1] have in their component a plane structure 2R, which is a basic structure, a kinematic chain plan 2-3 open.
In figure 1 can be monitored cinematic schema of the kinematic chain plan 2-3 open.

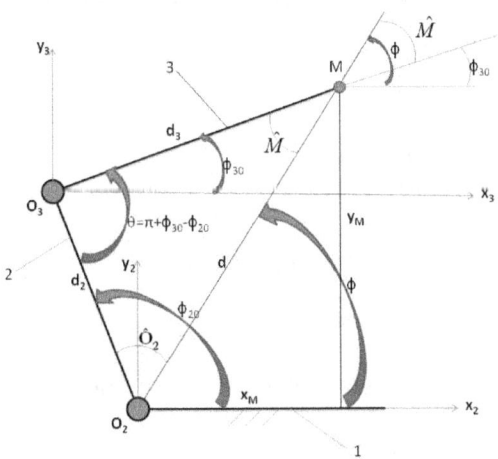

Fig. 1. *Cinematic schema of kinematic chain plan 2-3 linked to item 1 considered fixed*

In the reverse (inverse) kinematic, are known the kinematic parameters x_M and y_M, which represents the co-ordinates of scaling point M (endeffector M), and must be determined by analytical calculation the parameters $φ_{20}$ and $φ_{30}$ [11-15].

First, it determines the intermediary parameters, d and φ with relations (1).

$$\begin{cases} d^2 = x_M^2 + y_M^2; \quad d = \sqrt{x_M^2 + y_M^2} \\ \cos\varphi = \dfrac{x_M}{d} = \dfrac{x_M}{\sqrt{x_M^2 + y_M^2}}; \quad \sin\varphi = \dfrac{y_M}{d} = \dfrac{y_M}{\sqrt{x_M^2 + y_M^2}} \\ \varphi = sign(\sin\varphi) \cdot \arccos(\cos\varphi) \end{cases} \quad (1)$$

In some triangle O_2O_3M known lengths of the three sides, d_2, d_3 (constant) and d (variable), so that may be determined according to sides lengths all other elements of triangle, and more specifically its angles, trigonometric functions and their (us great interest in sin and cos).
For the purpose of determining angles can be used various methods (trigonometric, geometric, etc.), of which will be presented below one of them (as the most representative): trigonometric method.

Trigonometric method
Positions determination
Shall I write the equations of positions scaling (2) [11, 15]:

$$\begin{cases} d_2 \cdot \cos\varphi_{20} + d_3 \cdot \cos\varphi_{30} = x_M \\ d_2 \cdot \sin\varphi_{20} + d_3 \cdot \sin\varphi_{30} = y_M \\ \cos^2\varphi_{20} + \sin^2\varphi_{20} = 1 \\ \cos^2\varphi_{30} + \sin^2\varphi_{30} = 1 \end{cases} \quad (2)$$

These two equations problem scaling, trigonometric, with two unknown (φ_{20} *and* φ_{30}) is that they are more often met and slowly exceed (are equations trigonometric, transcendental equations, where unknown does not appear directly φ_{20} but also in the form $\cos\varphi_{20}$ and $\sin\varphi_{20}$, so that in reality within the framework of the two equations trigonometric don't have two unknown but four: $\cos\varphi_{20}$, $\sin\varphi_{20}$, $\cos\varphi_{30}$ and $\sin\varphi_{30}$). To resolve system we need two more equations, so that in the system (2) has been added two more equations trigonometric, more exactly the equations trigonometric of base "gold" as they say, for the angle φ_{20} and separately for the angle φ_{30}.
With a view to resolving the two equations of the system (2) shall be written in the form (3).

$$\begin{cases} d_2 \cdot \cos\varphi_{20} - x_M = -d_3 \cdot \cos\varphi_{30} \\ d_2 \cdot \sin\varphi_{20} - y_M = -d_3 \cdot \sin\varphi_{30} \end{cases} \quad (3)$$

Each equation of the system (3) rose square, after which summed up both equation (raised square) and obtain the equation of the form (4).

$$d_2^2 \cdot (\cos^2 \varphi_{20} + \sin^2 \varphi_{20}) + x_M^2 + y_M^2 - 2 \cdot d_2 \cdot x_M \cdot \cos \varphi_{20} - \\ - 2 \cdot d_2 \cdot y_M \cdot \sin \varphi_{20} = d_3^2 \cdot (\cos^2 \varphi_{30} + \sin^2 \varphi_{30})$$ (4)

Now is the time to use the two "equation of gold" trigonometric system written at the end (2), whereby the equation (4) becomes simplified form (5).

$$d_2^2 + x_M^2 + y_M^2 - 2 \cdot d_2 \cdot x_M \cdot \cos \varphi_{20} - 2 \cdot d_2 \cdot y_M \cdot \sin \varphi_{20} = d_3^2$$ (5)

Arrange the terms of this equation (5) in the form most convenient (6).

$$d_2^2 - d_3^2 + x_M^2 + y_M^2 = 2 \cdot d_2 \cdot (x_M \cdot \cos \varphi_{20} + y_M \cdot \sin \varphi_{20})$$ (6)

Divide the equation (6) with $2 \cdot d_2$ and will result in an new form (7).

$$x_M \cdot \cos \varphi_{20} + y_M \cdot \sin \varphi_{20} = \frac{d_2^2 - d_3^2 + x_M^2 + y_M^2}{2 \cdot d_2}$$ (7)

As shown in figure 1 shall be deducted and the relationship (8) which may also appears in the system (1).

$$x_M^2 + y_M^2 = d^2$$ (8)

Enter the expression (8) to (7) and multiply the fraction from the right with d, so that the expression (7) takes the form convenient (9).

$$x_M \cdot \cos \varphi_{20} + y_M \cdot \sin \varphi_{20} = \frac{d_2^2 + d^2 - d_3^2}{2 \cdot d_2 \cdot d} \cdot d$$ (9)

Now it's time of introduction of expression cosine of O_2 angle, depending on the sides of the some triangle $O_2 O_3 M$ (10).

$$\cos \hat{O}_2 = \frac{d_2^2 + d^2 - d_3^2}{2 \cdot d_2 \cdot d}$$ (10)

With the relation (10) the equation (9) becomes simplified form (11).

$$x_M \cdot \cos \varphi_{20} - d \cdot \cos \hat{O}_2 = -y_M \cdot \sin \varphi_{20} \quad (11)$$

We want to eliminate $\sin \varphi_{20}$, from which I isolate the term in sin, and got to his feet to square the equation (11), so that by using the equation gold (trigonometric) for the angle φ_{20} to transform sin in cos, equation become one of the second degree in $\cos \varphi_{20}$. After raising square (11) takes shape (12).

$$\begin{aligned} x_M^2 \cdot \cos^2 \varphi_{20} + d^2 \cdot \cos^2 \hat{O}_2 - 2 \cdot d \cdot x_M \cdot \cos \hat{O}_2 \cdot \cos \varphi_{20} = \\ = y_M^2 \cdot \sin^2 \varphi_{20} \end{aligned} \quad (12)$$

Using formula of gold, the expression (12) takes shape (13) which is arranges convenient by grouping terms, and bringing it to the form (14).

$$\begin{aligned} x_M^2 \cdot \cos^2 \varphi_{20} + d^2 \cdot \cos^2 \hat{O}_2 - 2 \cdot d \cdot x_M \cdot \cos \hat{O}_2 \cdot \cos \varphi_{20} = \\ = y_M^2 - y_M^2 \cdot \cos^2 \varphi_{20} \end{aligned} \quad (13)$$

$$\begin{aligned} (x_M^2 + y_M^2) \cdot \cos^2 \varphi_{20} - 2 \cdot d \cdot x_M \cdot \cos \hat{O}_2 \cdot \cos \varphi_{20} - \\ - (y_M^2 - d^2 \cdot \cos^2 \hat{O}_2) = 0 \end{aligned} \quad (14)$$

Discriminant of the equation (14) of second degree in cos obtained shall be calculated according to the relation (15).

$$\begin{cases} \Delta = d^2 \cdot x_M^2 \cdot \cos^2 \hat{O}_2 + d^2 \cdot (y_M^2 - d^2 \cdot \cos^2 \hat{O}_2) = \\ = d^2 \cdot (x_M^2 \cdot \cos^2 \hat{O}_2 + y_M^2 - d^2 \cdot \cos^2 \hat{O}_2) = \\ = d^2 \cdot (y_M^2 - y_M^2 \cdot \cos^2 \hat{O}_2) \\ = d^2 \cdot y_M^2 \cdot (1 - \cos^2 \hat{O}_2) = d^2 \cdot y_M^2 \cdot \sin^2 \hat{O}_2 \end{cases} \quad (15)$$

Radical from the second order of the discriminant is expressed as (16).

$$R = \sqrt{\Delta} = \sqrt{d^2 \cdot y_M^2 \cdot \sin^2 \hat{O}_2} = d \cdot y_M \cdot \sin \hat{O}_2 \quad (16)$$

Equation solutions (14) second degree in cos shall be written in the form (17).

$$\begin{cases} \cos\varphi_{20_{1,2}} = \dfrac{d \cdot x_M \cdot \cos\hat{O}_2 \mp d \cdot y_M \cdot \sin\hat{O}_2}{d^2} = \\ = \dfrac{x_M \cdot \cos\hat{O}_2 \mp y_M \cdot \sin\hat{O}_2}{d} = \\ = \dfrac{x_M}{d} \cdot \cos\hat{O}_2 \mp \dfrac{y_M}{d} \cdot \sin\hat{O}_2 \end{cases} \quad (17)$$

In the solutions (17) are replaced the reports with the corresponding functions trigonometric of the angle φ, expressions (17) mellowed form (18).

$$\begin{cases} \cos\varphi_{20_{1,2}} = \dfrac{x_M}{d} \cdot \cos\hat{O}_2 \mp \dfrac{y_M}{d} \cdot \sin\hat{O}_2 = \\ = \cos\varphi \cdot \cos\hat{O}_2 \mp \sin\varphi \cdot \sin\hat{O}_2 = \cos(\varphi \pm \hat{O}_2) \\ \\ \cos\varphi_{20} = \cos(\varphi \pm \hat{O}_2) \end{cases} \quad (18)$$

Now we're going back to the equation (11) that is ordered in the form (19), in order to solve them in sin. The equation (19) rose square and through the use of gold trigonometric equation of angle φ$_{20}$, is obtained form (20).

$$x_M \cdot \cos\varphi_{20} = d \cdot \cos\hat{O}_2 - y_M \cdot \sin\varphi_{20} \quad (19)$$

$$\begin{cases} x_M^2 \cdot \cos^2\varphi_{20} = d^2 \cdot \cos^2\hat{O}_2 + y_M^2 \cdot \sin^2\varphi_{20} - \\ -2 \cdot y_M \cdot d \cdot \cos\hat{O}_2 \cdot \sin\varphi_{20} \\ \\ x_M^2 - x_M^2 \cdot \sin^2\varphi_{20} = d^2 \cdot \cos^2\hat{O}_2 + y_M^2 \cdot \sin^2\varphi_{20} - \\ -2 \cdot y_M \cdot d \cdot \cos\hat{O}_2 \cdot \sin\varphi_{20} \\ \\ (x_M^2 + y_M^2) \cdot \sin^2\varphi_{20} - 2 \cdot y_M \cdot d \cdot \cos\hat{O}_2 \cdot \sin\varphi_{20} - \\ -(x_M^2 - d^2 \cdot \cos^2\hat{O}_2) = 0 \\ \\ d^2 \cdot \sin^2\varphi_{20} - 2 \cdot y_M \cdot d \cdot \cos\hat{O}_2 \cdot \sin\varphi_{20} - (x_M^2 - d^2 \cdot \cos^2\hat{O}_2) = 0 \end{cases} \quad (20)$$

Discriminant of the equation (20) of second degree in cos takes the form (21).

$$\begin{cases} \Delta = y_M^2 \cdot d^2 \cdot \cos^2 \hat{O}_2 + d^2 \cdot (x_M^2 - d^2 \cdot \cos^2 \hat{O}_2) = \\ = d^2 \cdot (x_M^2 + y_M^2 \cdot \cos^2 \hat{O}_2 - x_M^2 \cdot \cos^2 \hat{O}_2 - y_M^2 \cdot \cos^2 \hat{O}_2) = \\ = d^2 \cdot (x_M^2 - x_M^2 \cdot \cos^2 \hat{O}_2) = d^2 \cdot x_M^2 \cdot \sin^2 \hat{O}_2 \end{cases} \quad (21)$$

Equation Solutions (20) shall be written in the form (22).

$$\begin{cases} \sin \varphi_{20} = \dfrac{y_M \cdot d \cdot \cos \hat{O}_2 \pm x_M \cdot d \cdot \sin \hat{O}_2}{d^2} = \\ = \dfrac{y_M \cdot \cos \hat{O}_2 \pm x_M \cdot \sin \hat{O}_2}{d} = \dfrac{y_M}{d} \cdot \cos \hat{O}_2 \pm \dfrac{x_M}{d} \cdot \sin \hat{O}_2 = \\ = \sin \varphi \cdot \cos \hat{O}_2 \pm \cos \varphi \cdot \sin \hat{O}_2 = \sin(\varphi \pm \hat{O}_2) \\ \\ \sin \varphi_{20} = \sin(\varphi \pm \hat{O}_2) \end{cases} \quad (22)$$

Have been obtained relations (23), of which shall be deducted the basic relationship (24).

$$\begin{cases} \cos \varphi_{20} = \cos(\varphi \pm \hat{O}_2) \\ \sin \varphi_{20} = \sin(\varphi \pm \hat{O}_2) \end{cases} \quad (23)$$

$$\varphi_{20} = \varphi \pm \hat{O}_2 \quad (24)$$

Repeat the procedure and for determining φ_{30} angle, starting again from the system (2), in which the first two equations transcedental are rewritten in the form (25), with a view to eliminating φ_{20} angle at this time.

$$\begin{cases} d_2 \cdot \cos \varphi_{20} + d_3 \cdot \cos \varphi_{30} = x_M \\ d_2 \cdot \sin \varphi_{20} + d_3 \cdot \sin \varphi_{30} = y_M \\ \cos^2 \varphi_{20} + \sin^2 \varphi_{20} = 1 \\ \cos^2 \varphi_{30} + \sin^2 \varphi_{30} = 1 \end{cases} \quad (2)$$

$$\begin{cases} d_2 \cdot \cos \varphi_{20} = x_M - d_3 \cdot \cos \varphi_{30} \\ d_2 \cdot \sin \varphi_{20} = y_M - d_3 \cdot \sin \varphi_{30} \end{cases} \quad (25)$$

He stood up the two equations of the system (25) square and summed, resulting the equation of the form (26), which shall be made up in one of the forms easier (27) and (28).

$$d_2^2 = x_M^2 + y_M^2 + d_3^2 - 2 \cdot d_3 \cdot x_M \cdot \cos \varphi_{30} - 2 \cdot d_3 \cdot y_M \cdot \sin \varphi_{30} \quad (26)$$

$$x_M \cdot \cos \varphi_{30} + y_M \cdot \sin \varphi_{30} = d \cdot \frac{d^2 + d_3^2 - d_2^2}{2 \cdot d \cdot d_3} \quad (27)$$

$$x_M \cdot \cos \varphi_{30} + y_M \cdot \sin \varphi_{30} = d \cdot \cos \hat{M} \quad (28)$$

We would like to determine first the function cos, so we will isolate for the period began the term in sin, equation (28) putting it in the form (29), that by squaring generates the expression (30), which is arranges in the form (31).

$$x_M \cdot \cos \varphi_{30} - d \cdot \cos \hat{M} = -y_M \cdot \sin \varphi_{30} \quad (29)$$

$$x_M^2 \cdot \cos^2 \varphi_{30} + d^2 \cdot \cos^2 \hat{M} - 2 \cdot d \cdot x_M \cdot \cos \hat{M} \cdot \cos \varphi_{30} = \\ = y_M^2 - y_M^2 \cdot \cos^2 \varphi_{30} \quad (30)$$

$$d^2 \cdot \cos^2 \varphi_{30} - 2 \cdot d \cdot x_M \cdot \cos \hat{M} \cdot \cos \varphi_{30} - \\ - (y_M^2 - d^2 \cdot \cos^2 \hat{M}) = 0 \quad (31)$$

The equation (31) is an equation of degree II in cos, with the solutions given by the expression (32).

$$\begin{cases} \cos\varphi_{30} = \\ = \dfrac{d \cdot x_M \cdot \cos \hat{M} \pm \sqrt{d^2 \cdot x_M^2 \cdot \cos^2 \hat{M} + d^2 \cdot (y_M^2 - d^2 \cdot \cos^2 \hat{M})}}{d^2} = \\ = \dfrac{d \cdot x_M \cdot \cos \hat{M} \pm \sqrt{d^2 \cdot y_M^2 \cdot (1 - \cos^2 \hat{M})}}{d^2} = \\ = \dfrac{d \cdot x_M \cdot \cos \hat{M} \pm d \cdot y_M \cdot \sin \hat{M}}{d^2} = \\ = \dfrac{x_M}{d} \cdot \cos \hat{M} \pm \dfrac{y_M}{d} \cdot \sin \hat{M} = \cos\varphi \cdot \cos \hat{M} \pm \sin\varphi \cdot \sin \hat{M} = \\ = \cos(\varphi \mp \hat{M}) \\ \cos\varphi_{30} = \cos(\varphi \mp \hat{M}) \end{cases}$$

(32)

Write the following equation (28) in the form (33), where it is isolate from this time the term in cos with a view to abolishing it, so that it can be determined the sin.

$$x_M \cdot \cos\varphi_{30} = d \cdot \cos \hat{M} - y_M \cdot \sin\varphi_{30} \qquad (33)$$

The equation (33) rose square and obtain the equation of the form (34), which shall be made up in the form convenient (35).

$$\begin{aligned} & x_M^2 \cdot (1 - \sin^2\varphi_{30}) = \\ & = d^2 \cdot \cos^2 \hat{M} + y_M^2 \cdot \sin^2\varphi_{30} - 2 \cdot y_M \cdot d \cdot \cos \hat{M} \cdot \sin\varphi_{30} \end{aligned} \qquad (34)$$

$$\begin{aligned} & d^2 \cdot \sin^2\varphi_{30} - 2 \cdot y_M \cdot d \cdot \cos \hat{M} \cdot \sin\varphi_{30} - \\ & - (x_M^2 - d^2 \cdot \cos^2 \hat{M}) = 0 \end{aligned} \qquad (35)$$

The expression (35) is an equation of degree II in sin, which allowed some solutions given by the relationship (36).

$$\begin{cases} \sin\varphi_{30} = \\ = \dfrac{d \cdot y_M \cdot \cos\hat{M} \mp \sqrt{d^2 \cdot y_M^2 \cdot \cos^2\hat{M} + d^2 \cdot (x_M^2 - d^2 \cdot \cos^2\hat{M})}}{d^2} = \\ = \dfrac{d \cdot y_M \cdot \cos\hat{M} \mp \sqrt{d^2 \cdot x_M^2 \cdot (1 - \cos^2\hat{M})}}{d^2} = \\ = \dfrac{d \cdot y_M \cdot \cos\hat{M} \mp d \cdot x_M \cdot \sin\hat{M}}{d^2} = \\ = \dfrac{y_M}{d} \cdot \cos\hat{M} \mp \dfrac{x_M}{d} \cdot \sin\hat{M} = \sin\varphi \cdot \cos\hat{M} \mp \cos\varphi \cdot \sin\hat{M} = \\ = \sin(\varphi \mp \hat{M}) \\ \sin\varphi_{30} = \sin(\varphi \mp \hat{M}) \end{cases}$$

(36)

Are retained relations (37) of which shall be deducted and the expression (38).

$$\begin{cases} \cos\varphi_{30} = \cos(\varphi \mp \hat{M}) \\ \sin\varphi_{30} = \sin(\varphi \mp \hat{M}) \end{cases} \quad (37)$$

$$\varphi_{30} = \varphi \mp \hat{M} \quad (38)$$

Determining the velocities

In the system (1) derived be retained only relations (39), necessary in the study of the inverse kinematics of system. Start at the relationship linking cosine $\hat{O}_2$ angle to the sides of the triangle, which is derived with respect to time, and is given the value $\dot{\hat{O}}_2$, writing simpler $\dot{O}_2$ (relations 40).

$$\begin{cases} \dot{\varphi} = \dfrac{\dot{y}_M \cdot \cos\varphi - \dot{x}_M \cdot \sin\varphi}{d} \\ \dot{d} = \dfrac{x_M \cdot \dot{x}_M + y_M \cdot \dot{y}_M}{d} \end{cases} \quad (39)$$

$$\begin{cases} 2 \cdot d_2 \cdot d \cdot \cos O_2 = d_2^2 - d_3^2 + d^2 \\ 2 \cdot d_2 \cdot \dot{d} \cdot \cos O_2 - 2 \cdot d_2 \cdot d \cdot \sin O_2 \cdot \dot{O}_2 = 2 \cdot d \cdot \dot{d} \Rightarrow \\ \Rightarrow \dot{O}_2 = \dfrac{d_2 \cdot \dot{d} \cdot \cos O_2 - d \cdot \dot{d}}{d_2 \cdot d \cdot \sin O_2} \end{cases} \quad (40)$$

Is derived relationship (24) and obtain angular speed (the relationship 41).

$$\varphi_{20} = \varphi \pm \hat{O}_2 \quad (24)$$

$$\omega_{20} \equiv \dot{\varphi}_{20} = \dot{\varphi} \pm \dot{O}_2 \quad (41)$$

To determine ω_{20} (the relationship 41) we need $\dot{\varphi}$ which is calculated from (39), and $\dot{O}_2$ which is to be determined from (40). In turn $\dot{O}_2$ requires for its calculation $\dot{d}$ who is still determined from the system (39) as well.

Input speeds $\dot{x}_M$ and $\dot{y}_M$ are known, are imposed in input data, or choose convenient, times can be calculated on the basis of such criteria as required.
Similarly shall be determined and angular speed $\omega_{30} \equiv \dot{\varphi}_{30}$.

$$\begin{cases} 2 \cdot d_3 \cdot d \cdot \cos M = d_3^2 - d_2^2 + d^2 \\ 2 \cdot d_3 \cdot \dot{d} \cdot \cos M - 2 \cdot d_3 \cdot d \cdot \sin M \cdot \dot{M} = 2 \cdot d \cdot \dot{d} \Rightarrow \\ \Rightarrow \dot{M} = \dfrac{d_3 \cdot \dot{d} \cdot \cos M - d \cdot \dot{d}}{d_3 \cdot d \cdot \sin M} \end{cases} \quad (42)$$

It derived relationship (38) in order to obtain angular speed $\omega_{30} \equiv \dot{\varphi}_{30}$, (the expression 43). $\dot{\varphi}$ shall be calculated with the expression already known from the system (39), and $\dot{M}$ shall be determined from the system (42) and with the aid of (39) that it determines and on the $\dot{d}$.

$$\varphi_{30} = \varphi \mp \hat{M} \quad (38)$$

$$\omega_{30} \equiv \dot{\varphi}_{30} = \dot{\varphi} \mp \dot{M} \qquad (43)$$

Determining the accelerations

From the system (39) derived, are obtained relations (44), required in the study of the system accelerations in inverse kinematic. The relationship of the system (40) is derived a second time with time, and obtain the system (45).

$$\begin{cases} \ddot{\varphi} = \dfrac{\ddot{y}_M \cdot \cos\varphi - \ddot{x}_M \cdot \sin\varphi - \dot{y}_M \cdot \sin\varphi \cdot \dot{\varphi} - \dot{x}_M \cdot \cos\varphi \cdot \dot{\varphi} - \dot{d} \cdot \dot{\varphi}}{d} \\ \ddot{d} = \dfrac{\dot{x}_M^2 + x_M \cdot \ddot{x}_M + \dot{y}_M^2 + y_M \cdot \ddot{y}_M - \dot{d}^2}{d} \end{cases} \qquad (44)$$

$$\begin{cases} 2 \cdot d_2 \cdot d \cdot \cos O_2 = d_2^2 - d_3^2 + d^2 \\ \\ 2 \cdot d_2 \cdot \dot{d} \cdot \cos O_2 - 2 \cdot d_2 \cdot d \cdot \sin O_2 \cdot \dot{O}_2 = 2 \cdot d \cdot \dot{d} \Rightarrow \\ \Rightarrow d_2 \cdot d \cdot \sin O_2 \cdot \dot{O}_2 = d_2 \cdot \dot{d} \cdot \cos O_2 - d \cdot \dot{d} \\ \\ \ddot{O}_2 = \dfrac{\ddot{d}d_2 \cos O_2 - \ddot{d}d - 2\dot{d}d_2 \sin O_2 \cdot \dot{O}_2 - dd_2 \cos O_2 \cdot \dot{O}_2^2 - \dot{d}^2}{d_2 \cdot d \cdot \sin O_2} \end{cases} \qquad (45)$$

The following is derived expression (41) and obtain the relationship (46), which generates an absolute angular acceleration $\varepsilon_2 \equiv \varepsilon_{20}$, which can be calculated with $\ddot{\varphi}$ deducted from the system (44), and with $\ddot{O}_2$ obtained from the system (45), and for its determination ($\ddot{O}_2$ determination) it is necessary and $\ddot{d}$ calculated all from (44).

$$\omega_{20} \equiv \dot{\varphi}_{20} = \dot{\varphi} \pm \dot{O}_2 \qquad (41)$$

$$\varepsilon_2 \equiv \varepsilon_{20} = \dot{\omega}_{20} \equiv \ddot{\varphi}_{20} = \ddot{\varphi} \pm \ddot{O}_2 \qquad (46)$$

Now derived a second time expression (42) and obtain the system (47).

$$\begin{cases} 2 \cdot d_3 \cdot d \cdot \cos M = d_3^2 - d_2^2 + d^2 \\[2mm] 2 \cdot d_3 \cdot \dot{d} \cdot \cos M - 2 \cdot d_3 \cdot d \cdot \sin M \cdot \dot{M} = 2 \cdot d \cdot \dot{d} \Rightarrow \\ \Rightarrow d_3 \cdot d \cdot \sin M \cdot \dot{M} = d_3 \cdot \dot{d} \cdot \cos M - d \cdot \dot{d} \hfill (47) \\[2mm] \ddot{M} = \dfrac{\ddot{d}d_3 \cos M - \ddot{d}d - 2\dot{d}d_3 \sin M \cdot \dot{M} - dd_3 \cos M \cdot \dot{M}^2 - \dot{d}^2}{d_3 \cdot d \cdot \sin M} \end{cases}$$

Is derived again with time relationship (43), and obtain the expression (64) of the absolute angular acceleration $\varepsilon_3 \equiv \varepsilon_{30}$ which is to be determined with $\ddot{\varphi}$ and $\ddot{M}$.

$\ddot{\varphi}$ is obtained from the system (44), and $\ddot{M}$ shall be calcullated from the system (47), and also needs $\ddot{d}$ obtain all of the system (44).

$$\omega_{30} \equiv \dot{\varphi}_{30} = \dot{\varphi} \mp \dot{M} \hfill (43)$$

$$\varepsilon_3 \equiv \varepsilon_{30} = \dot{\omega}_{30} \equiv \ddot{\varphi}_{30} = \ddot{\varphi} \mp \ddot{M} \hfill (48)$$

Conclussions

Anthropomorphic robots have in their component a plane structure 2R, which is a basic structure. In the reverse (inverse) kinematic, are known the kinematic parameters x_M and y_M, which represents the co-ordinates of scaling point M (endeffector M), and must be determined by analytical calculation the parameters φ_{20} and φ_{30}. For the purpose of determining angles can be used various methods (trigonometric, geometric, etc.), of which was presented the trigonometric method, which is a direct and rapid method.

This method has and the great advantage to be more easier to understand. With it you can avoid spatial methods of calculation.

The method proposed in this work has the advantage of greatly simplify day-to-systems design calculations to mechatronics and robotic. Win time, saves work, it's possible to work out a direct synthesis easier serial systems, without it being necessary experimental testing. For teaching method is a total simplification of the design, and thus increasing the understanding of this phenomenon.

References

1. Nelson Diaz Aldana, César Leonardo Trujillo, José Guillermo Guarnizo, Active and reactive power flow regulation for a grid connected vsc based on fuzzy controllers, Revista Facultad de Ingeniería, No. 66 (2013), pp. 118-130, ISSN: 0120-6230. Available from:

    http://aprendeenlinea.udea.edu.co/revistas/index.php/ingenieria/article/view/15229/13232

2. Dong, H., Giakoumidis, N., Figueroa, N., and Mavridis, N., (2013). Approaching Behaviour Monitor and Vibration Indication in Developing a General Moving Object Alarm System (GMOAS), International Journal of Advanced Robotic Systems, Hanafiah Yussof (Ed.), ISBN: 1729-8806, InTech, DOI: 10.5772/56586. Available from:

    http://www.intechopen.com/journals/international_journal_of_advanced_robotic_systems/approaching-behaviour-monitor-and-vibration-indication-in-developing-a-general-moving-object-alarm-s.

3. Flavio de Melo, L., Reis Alves, S.F., Rosário, J.M., Mobile Robot Navigation Modelling, Control and Applications, in International Review on Modelling and Simulations, April 2012, Vol. 5, N. 2B, pp. 1059-1068. Available from:

    http://www.praiseworthyprize.com/IREMOS-latest/IREMOS_vol_5_n_2.html.

4. Lee, B.J., (2013). Geometrical Derivation of Differential Kinematics to Calibrate Model Parameters of Flexible Manipulator, International Journal of Advanced Robotic Systems, Jaime Gallardo-Alvarado, Ramon Rodrıguez-Castro (Ed.), ISBN: 1729-8806, InTech, DOI: 10.5772/55592. Available from:
http://www.intechopen.com/journals/international_journal_of_advanced_robotic_systems/geometrical-derivation-of-differential-kinematics-to-calibrate-model-parameters-of-flexible-manipula.

5. Garcia, E.; Jimenez, M.A.; De Santos, P.G.; Armada, M., "The evolution of robotics research," Robotics & Automation Magazine, IEEE , vol.14, no.1, pp.90,103, March 2007. Available from:

    http://ieeexplore.ieee.org/stamp/stamp.jsp?tp=&arnumber=4141037&isnumber=4141014.

6. He, B., Wang, Z., Li, Q., Xie, H., and Shen, R., (2013). An Analytic Method for the Kinematics and Dynamics of a Multiple-Backbone Continuum Robot, IJARS, Patricia Melin (Ed.), ISBN: 1729-8806, InTech, DOI: 10.5772/54051. Available from:

    http://www.intechopen.com/journals/international_journal_of_advanced_robotic_systems/an-analytic-method-for-the-kinematics-and-dynamics-of-a-multiple-backbone-continuum-robot.

7. Liu, H., Zhou, W., Lai, X., and Zhu, S., (2013). An Efficient Inverse Kinematic Algorithm for a PUMA560-Structured Robot Manipulator, IJARS, Jaime Gallardo-Alvarado, Ramon Rodrıguez-Castro (Ed.), ISBN: 1729-8806, InTech, DOI: 10.5772/56403. Available from:

http://www.intechopen.com/journals/international_journal_of_advanced_robotic_systems/an-efficient-inverse-kinematic-algorithm-for-a-puma560-structured-robot-manipulator.

8. Garcia-Murillo, M., Gallardo-Alvarado, J., and Castillo-Castaneda, E., (2013). Finding the Generalized Forces of a Series-Parallel Manipulator, IJARS, Jaime Gallardo-Alvarado, Ramon Rodriguez-Castro (Ed.), ISBN: 1729-8806, InTech, DOI: 10.5772/53824. Available from:

http://www.intechopen.com/journals/international_journal_of_advanced_robotic_systems/finding-the-generalized-forces-of-a-series-parallel-manipulator.

9. Padula, F., and Perdereau, V., (2013). An On-Line Path Planner for Industrial Manipulators, International Journal of Advanced Robotic Systems, Antonio Visioli (Ed.), ISBN: 1729-8806, InTech, DOI: 10.5772/55063. Available from:

http://www.intechopen.com/journals/international_journal_of_advanced_robotic_systems/an-on-line-path-planner-for-industrial-manipulators.

10. Perumaal, S., and Jawahar, N., (2013). Automated Trajectory Planner of Industrial Robot for Pick-and-Place Task, IJARS, Antonio Visioli (Ed.), ISBN: 1729-8806, InTech, DOI: 10.5772/53940. Available from:

http://www.intechopen.com/journals/international_journal_of_advanced_robotic_systems/automated-trajectory-planner-of-industrial-robot-for-pick-and-place-task.

11. Petrescu, F.I., Petrescu, R.V., Cinematics of the 3R Dyad, in journal Engevista, vol. 15, No. 2, (2013), August 2013, ISSN 1415-7314, pp. 118-124. Available from:

http://www.uff.br/engevista/seer/index.php/engevista/article/view/376.

12. Petrescu, F.I., Petrescu, R.V., Kinematics of the Planar Quadrilateral Mechanism, in journal Engevista, vol. 14, No. 3, (2012), December 2012, ISSN 1415-7314, pp. 345-348. Available from:

http://www.uff.br/engevista/seer/index.php/engevista/article/view/377.

13. Petrescu, F.I., Petrescu, R.V., Mecatronica – Sisteme Seriale si Paralele, Create Space publisher, USA, March 2012, ISBN 978-1-4750-6613-5, 128 pages, Romanian edition.

14. Petrescu, F.I, Petrescu, R.V, Mechanical Systems, Serial and Parallel – Course (in romanian), LULU Publisher, London, UK, February 2011, 124 pages, ISBN 978-1-4466-0039-9, Romanian edition.

15. Petrescu, F.I., Grecu, B., Comanescu, A., Petrescu, R.V., Some Mechanical Design Elements. In the 3rd International Conference on Computational Mechanics and Virtual Engineering, COMEC 2009, Brașov, October 2009, ISBN 978-973-598-572-1, Edit. UTB, pp. 520-525.

16. Reddy, P., Shihabudheen K.V., Jacob, J., Precise Non Linear Modeling of Flexible Link Flexible Joint Manipulator, in International Review on Modelling and Simulations, June 2012, Vol. 5, N. 3B, pp. 1368-1374. Available from:

http://www.praiseworthyprize.com/IREMOS-latest/IREMOS_vol_5_n_3.html.

17. Tang, X., Sun, D., and Shao, Z., (2013). The Structure and Dimensional Design of a Reconfigurable PKM, IJARS, Sumeet S Aphale (Ed.), ISBN: 1729-8806, InTech, DOI: 10.5772/54696. Available from:

    http://www.intechopen.com/journals/international_journal_of_advanced_robotic_systems/the-structure-and-dimensional-design-of-a-reconfigurable-pkm.

18. Tong, G., Gu, J., and Xie, W., (2013). Virtual Entity-Based Rapid Prototype for Design and Simulation of Humanoid Robots, International Journal of Advanced Robotic Systems, Guangming Xie (Ed.), ISBN: 1729-8806, InTech, DOI: 10.5772/55936. Available from:

    http://www.intechopen.com/journals/international_journal_of_advanced_robotic_systems/virtual-entity-based-rapid-prototype-for-design-and-simulation-of-humanoid-robots.

19. Wang, K., Luo, M., Mei, T., Zhao, J., and Cao, Y., (2013). Dynamics Analysis of a Three-DOF Planar Serial-Parallel Mechanism for Active Dynamic Balancing with Respect to a Given Trajectory, International Journal of Advanced Robotic Systems, Sumeet S Aphale (Ed.), ISBN: 1729-8806, InTech, DOI: 10.5772/54201. Available from:

    http://www.intechopen.com/journals/international_journal_of_advanced_robotic_systems/dynamics-analysis-of-a-three-dof-planar-serial-parallel-mechanism-for-active-dynamic-balancing-with-

20. Wen, S., Zhu, J., Li, X., Rad, A., Chen, X., (2012). End-Point Contact Force Control with Quantitative Feedback Theory for Mobile Robots, IJARS, Houxiang Zhang, Shengyong Chen (Ed.), ISBN: 1729-8806, InTech, DOI: 10.5772/53742. Available from:

    http://www.intechopen.com/journals/international_journal_of_advanced_robotic_systems/end-point-contact-force-control-with-quantitative-feedback-theory-for-mobile-robots.

Chapter 06_The Transition from 2R System to the 3R System

In Figure 1 is illustrated spatial kinematic chain.

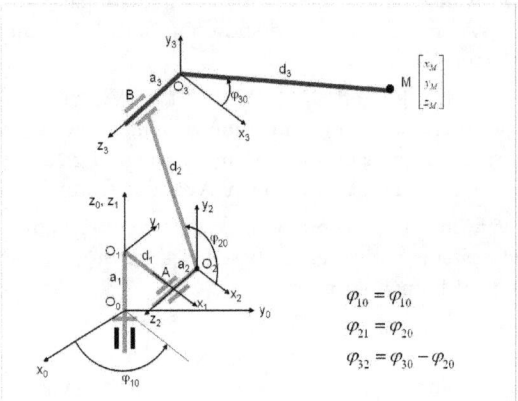

Fig. 1. *Geometro spatial kinematic Diagram of a modern structures 3R (anthropomorphic)*

The will make the transition from flat to the spatial movement.

Dimensions plane x_2Oy_2 will project to the axs $zO\rho$. So the length in the vertical axis flat Oy will be projected on the vertical axis spatial Oz by adding constant a_1, and the length of the horizontal axis flat Ox will be projected on the horizontal spatial axis $O\rho$ by adding constant d_1, in accordance with the data relations system (1).

$$\begin{cases} \rho_{M'} = d_1 + x_M^P \\ z_M = a_1 + y_M^P \end{cases} \quad (1)$$

Point projections M on flat axs shall be marked with the index higher P (plan), in order to detect the presence of appropriate spatial axs.

Due to the fact that the plan of vertical projection axis is moved away from $O\rho$ with a distance constant a_2+a_3, (the plan of vertically work is not projected directly on the axis $O\rho$, but on an axis parallel to it away with a length a_2+a_3), projection of point M on horizontal plane in space does not fall in M' but at the point M" (see figure 1).

Because of this projections of M on the spatial axs Ox and Oy, will not be those of point M' but of the point M", in accordance with the data relations system (2).

$$\begin{cases} x_M = \rho_{M'} \cdot \cos\varphi_{10} + (a_2 + a_3) \cdot \cos\left(\varphi_{10} + \dfrac{\pi}{2}\right) \\ y_M = \rho_{M'} \cdot \sin\varphi_{10} + (a_2 + a_3) \cdot \sin\left(\varphi_{10} + \dfrac{\pi}{2}\right) \end{cases} \quad (2)$$

We would like to eliminate the angle of 90 deg of our relations (2), which have had an important role in the understanding explanatory phenomenon, in order to see how to write the equations for switching from the axs plane in the space, being here (in the horizontal plane in space) comes to a rotation, whose relationships are not to be retained automatically, but deducted logic, that was the reason for which we pass immediately from the system determined logic (2) convenient system (3), which would have been obtained from (2) by eliminating angle of 90 deg from the trigonometric relations.

$$\begin{cases} x_M = \rho_{M'} \cdot \cos\varphi_{10} - (a_2 + a_3) \cdot \sin\varphi_{10} \\ y_M = \rho_{M'} \cdot \sin\varphi_{10} + (a_2 + a_3) \cdot \cos\varphi_{10} \end{cases} \quad (3)$$

Perhaps it may seem a little difficult method used, but compared to the methods matrix space, it is extremely simple and straightforward, helping to transform spatial movement in a single flat surface, much easier to understand and studied. The system (4) centralizes all relationships move from flat to the spatial movement.

$$\begin{cases} x_M = \left(d_1 + x_M^P\right) \cdot \cos\varphi_{10} - (a_2 + a_3) \cdot \sin\varphi_{10} \\ y_M = \left(d_1 + x_M^P\right) \cdot \sin\varphi_{10} + (a_2 + a_3) \cdot \cos\varphi_{10} \\ z_M = a_1 + y_M^P \end{cases} \quad (4)$$

By replacing in (4) the values of x_M^P and y_M^P obtain the equations absolute space (system 5).

$$\begin{cases} x_M = \left(d_1 + d_2 \cdot \cos\varphi_{20} + d_3 \cdot \cos\varphi_{30}\right) \cdot \\ \quad \cdot \cos\varphi_{10} - (a_2 + a_3) \cdot \sin\varphi_{10} \\ \\ y_M = \left(d_1 + d_2 \cdot \cos\varphi_{20} + d_3 \cdot \cos\varphi_{30}\right) \cdot \\ \quad \cdot \sin\varphi_{10} + (a_2 + a_3) \cdot \cos\varphi_{10} \\ \\ z_M = a_1 + d_2 \cdot \sin\varphi_{20} + d_3 \cdot \sin\varphi_{30} \end{cases} \quad (5)$$

For the purpose of determining easier the velocities and acceleration in the system (4), it is noted $a_2 + a_3$ with a, so that (4) achieve the stature (6) simplified scheme.

$$\begin{cases} x_M = (d_1 + x_M^P) \cdot \cos\varphi_{10} - a \cdot \sin\varphi_{10} \\ y_M = (d_1 + x_M^P) \cdot \sin\varphi_{10} + a \cdot \cos\varphi_{10} \\ z_M = a_1 + y_M^P \end{cases} \quad (6)$$

It is derived as a function of time the system of spatial positions (6) and obtains the system spatial of velocities (7).

$$\begin{cases} \dot{x}_M = \dot{x}_M^P \cdot \cos\varphi_{10} - (d_1 + x_M^P) \cdot \\ \quad \cdot \sin\varphi_{10} \cdot \dot{\varphi}_{10} - a \cdot \cos\varphi_{10} \cdot \dot{\varphi}_{10} \\ \\ \dot{y}_M = \dot{x}_M^P \cdot \sin\varphi_{10} + (d_1 + x_M^P) \cdot \\ \quad \cdot \cos\varphi_{10} \cdot \dot{\varphi}_{10} - a \cdot \sin\varphi_{10} \cdot \dot{\varphi}_{10} \\ \\ \dot{z}_M = \dot{y}_M^P \end{cases} \quad (7)$$

Is derived as a function of time the spatial velocities system (41) and obtains the spatial system of accelerations (8), which is restricted to the form (9).

$$\begin{cases} \ddot{x}_M = \ddot{x}_M^P \cdot \cos\varphi_{10} - \dot{x}_M^P \cdot \sin\varphi_{10} \cdot \dot{\varphi}_{10} - \\ \quad - \dot{x}_M^P \cdot \sin\varphi_{10} \cdot \dot{\varphi}_{10} - \\ \quad - (d_1 + x_M^P) \cdot \cos\varphi_{10} \cdot \dot{\varphi}_{10}^2 + a \cdot \sin\varphi_{10} \cdot \dot{\varphi}_{10}^2 \\ \ddot{y}_M = \ddot{x}_M^P \cdot \sin\varphi_{10} + \dot{x}_M^P \cdot \cos\varphi_{10} \cdot \dot{\varphi}_{10} + \\ \quad + \dot{x}_M^P \cdot \cos\varphi_{10} \cdot \dot{\varphi}_{10} - \\ \quad - (d_1 + x_M^P) \cdot \sin\varphi_{10} \cdot \dot{\varphi}_{10}^2 - a \cdot \cos\varphi_{10} \cdot \dot{\varphi}_{10}^2 \\ \\ \ddot{z}_M = \ddot{y}_M^P \end{cases} \quad (8)$$

$$\begin{cases} \ddot{x}_M = \left[\ddot{x}_M^P - (d_1 + x_M^P)\cdot \dot{\varphi}_{10}^2\right]\cdot \cos\varphi_{10} - \\ \quad - (2\cdot \dot{x}_M^P - a\cdot \dot{\varphi}_{10})\cdot \dot{\varphi}_{10}\cdot \sin\varphi_{10} \\ \\ \ddot{y}_M = \left[\ddot{x}_M^P - (d_1 + x_M^P)\cdot \dot{\varphi}_{10}^2\right]\cdot \sin\varphi_{10} + \\ \quad + (2\cdot \dot{x}_M^P - a\cdot \dot{\varphi}_{10})\cdot \dot{\varphi}_{10}\cdot \cos\varphi_{10} \\ \\ \ddot{z}_M = \ddot{y}_M^P \end{cases} \quad (9)$$

The system spatial of velocities (7) is restricted to the form (10), which by using numbers awarded u and v shall be simplified and be written in the form (11). And the system of accelerations (9) can be restricted to form (12), with notations w, t.

$$\begin{cases} \dot{x}_M = (\dot{x}_M^P - a\cdot \dot{\varphi}_{10})\cdot \cos\varphi_{10} - \\ \quad - (d_1 + x_M^P)\cdot \dot{\varphi}_{10}\cdot \sin\varphi_{10} \\ \\ \dot{y}_M = (\dot{x}_M^P - a\cdot \dot{\varphi}_{10})\cdot \sin\varphi_{10} + \\ \quad + (d_1 + x_M^P)\cdot \dot{\varphi}_{10}\cdot \cos\varphi_{10} \\ \\ \dot{z}_M = \dot{y}_M^P \end{cases} \quad (10)$$

$$\begin{cases} \dot{x}_M = u\cdot \cos\varphi_{10} - v\cdot \sin\varphi_{10} \\ \dot{y}_M = u\cdot \sin\varphi_{10} + v\cdot \cos\varphi_{10} \\ \dot{z}_M = \dot{y}_M^P \\ \\ u = \dot{x}_M^P - a\cdot \dot{\varphi}_{10}; \quad v = (d_1 + x_M^P)\cdot \dot{\varphi}_{10} \end{cases} \quad (11)$$

$$\begin{cases} \ddot{x}_M = w \cdot \cos\varphi_{10} - t \cdot \sin\varphi_{10} \\ \ddot{y}_M = w \cdot \sin\varphi_{10} + t \cdot \cos\varphi_{10} \\ \ddot{z}_M = \ddot{y}_M^P \\ \\ w = \ddot{x}_M^P - (d_1 + x_M^P) \cdot \dot{\varphi}_{10}^2; \\ t = (2 \cdot \dot{x}_M^P - a \cdot \dot{\varphi}_{10}) \cdot \dot{\varphi}_{10} \end{cases} \quad (12)$$

Further, it will present positions, speeds, and spatial accelerations, written all restricted within the framework of the system (13).

$$\begin{cases} Positions: \\ x_M = s \cdot \cos\varphi_{10} - a \cdot \sin\varphi_{10} \\ y_M = s \cdot \sin\varphi_{10} + a \cdot \cos\varphi_{10} \\ z_M = a_1 + y_M^P \\ cu \quad s = d_1 + x_M^P; \quad a = a_2 + a_3 \\ \\ Velocities: \\ \dot{x}_M = u \cdot \cos\varphi_{10} - v \cdot \sin\varphi_{10} \\ \dot{y}_M = u \cdot \sin\varphi_{10} + v \cdot \cos\varphi_{10} \\ \dot{z}_M = \dot{y}_M^P \\ cu \quad u = \dot{x}_M^P - a \cdot \dot{\varphi}_{10}; \quad v = (d_1 + x_M^P) \cdot \dot{\varphi}_{10} \\ \\ Accelerations: \\ \ddot{x}_M = w \cdot \cos\varphi_{10} - t \cdot \sin\varphi_{10} \\ \ddot{y}_M = w \cdot \sin\varphi_{10} + t \cdot \cos\varphi_{10} \\ \ddot{z}_M = \ddot{y}_M^P \\ \\ with: \\ w = \ddot{x}_M^P - (d_1 + x_M^P) \cdot \dot{\varphi}_{10}^2; \\ t = (2 \cdot \dot{x}_M^P - a \cdot \dot{\varphi}_{10}) \cdot \dot{\varphi}_{10} \end{cases} \quad (13)$$

Vector module position point spatial end effector M, in cartesian space-fixed system is given by the relation (14).

$$r_M = \sqrt{x_M^2 + y_M^2 + z_M^2} =$$
$$= \sqrt{s^2 + a^2 + (a_1 + y_M^P)^2} \qquad (14)$$

Vector module absolute velocity point spatial end effector M, in cartesian space-fixed system is given by the relation (15).

$$v_M = \sqrt{\dot{x}_M^2 + \dot{y}_M^2 + \dot{z}_M^2} =$$
$$= \sqrt{u^2 + v^2 + \dot{y}_M^{P\,2}} \qquad (15)$$

Vector module absolute acceleration point spatial end effector M, in cartesian space-fixed system is given by the relation (16).

$$a_M = \sqrt{\ddot{x}_M^2 + \ddot{y}_M^2 + \ddot{z}_M^2} =$$
$$= \sqrt{w^2 + t^2 + \ddot{y}_M^{P\,2}} \qquad (16)$$

In the system (17) provide a review of the three parameters space of absolute point M: absolute position, absolute velocity, acceleration absolute.

$$\begin{cases} r_M = \sqrt{x_M^2 + y_M^2 + z_M^2} = \\ \quad = \sqrt{s^2 + a^2 + (a_1 + y_M^P)^2} \\ \\ v_M = \sqrt{\dot{x}_M^2 + \dot{y}_M^2 + \dot{z}_M^2} = \\ \quad = \sqrt{u^2 + v^2 + \dot{y}_M^{P\,2}} \\ \\ a_M = \sqrt{\ddot{x}_M^2 + \ddot{y}_M^2 + \ddot{z}_M^2} = \\ \quad = \sqrt{w^2 + t^2 + \ddot{y}_M^{P\,2}} \end{cases} \qquad (17)$$

Chapter 07_ Kinetostatic and Dynamics

1. INTRODUCTION

In this chapter we present a method for determining the kinetostatic parameters to a 3R dyad (see the Figure 1) [1-4].

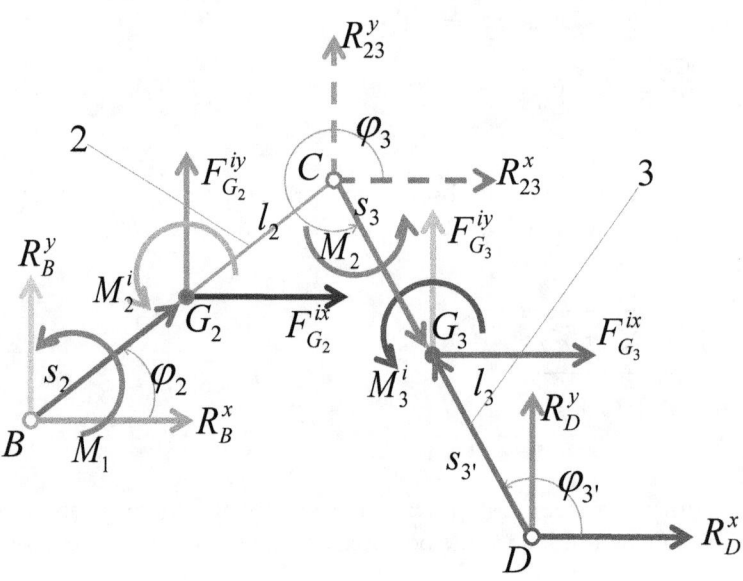

Fig. 1. *The kinetostatic parameters to a 3R dyad*

To generalize the method and to the 2R robots, are introduced and the two moments M1, M2. This 2R module, is the principal from the android rotation robotic structures and mechatronic structures [1], [3].

The 3R dyad has two elements, noted with 2 and 3. Their lengths are l2 and l3.

If the 3R dyad is coupling to a 4R mechanism, we note the forces which give the entry into dyad, with R12 and R03. In case the structure 2-3 is using to a robot or to another mechanism, we note the entrance forces, with RB and RD.

One proposes to determine the forces from joints: RB, RD, R23.

Figure 1 shows a schematic diagram of the 3R dyad minimum kinetostatic (loaded with the inertia forces, considered external forces).

For if there are additional external forces, such as technological resistances will be added as well.

One can consider and the forces of gravity, if mechanism operates strictly vertically and working speeds are low [1-4].

2. DETERMINING THE FORCES FROM JOINTS

The joints forces represent the interior loads (internal forces).

One proposes to determine these (internal) forces.

We start with the internal force R$_B$, which is divided in two components in a cartesian planar system: R_B^x, R_B^y.

If external forces are known in general (are given, determined, calculated), internal forces (reactions of kinematic couplings) results from the balance of forces and moments of the dyad [2], [3], [4].

To start [3] we are writing an equation representing the sum of the moments from element 2 in relation to the point C, and another relationship which represent the sum of all moments from entire dyad, in relation to the point D (system 1).

$$\begin{cases} \sum M_C^{(2)} = 0 \Rightarrow R_B^x \cdot (y_C - y_B) - \\ R_B^y \cdot (x_C - x_B) + M_1 + \\ + F_{G_2}^{ix} \cdot (y_C - y_{G_2}) - \\ - F_{G_2}^{iy} \cdot (x_C - x_{G_2}) + M_2^i = 0 \\ \sum M_D^{(2,3)} = 0 \Rightarrow R_B^x \cdot (y_D - y_B) - \\ - R_B^y \cdot (x_D - x_B) + M_1 + \\ + F_{G_2}^{ix} \cdot (y_D - y_{G_2}) - M_2^i + M_2 + \\ - F_{G_2}^{iy} \cdot (x_D - x_{G_2}) + M_3^i + \\ + F_{G_3}^{ix} \cdot (y_D - y_{G_3}) - \\ - F_{G_3}^{iy} \cdot (x_D - x_{G_3}) = 0 \end{cases} \quad (1)$$

The two equations are rewritten in the form of the system (2).

$$\begin{cases}
(y_C - y_B) \cdot R_B^x - (x_C - x_B) \cdot R_B^y = -M_1 - \\
- F_{G_2}^{ix} \cdot (y_C - y_{G_2}) + F_{G_2}^{iy} \cdot (x_C - x_{G_2}) - M_2^i \\
(y_D - y_B) \cdot R_B^x - (x_D - x_B) \cdot R_B^y = \\
= -M_1 - F_{G_2}^{ix} \cdot (y_D - y_{G_2}) + \\
+ F_{G_2}^{iy} \cdot (x_D - x_{G_2}) - M_2^i - M_2 - \\
- F_{G_3}^{ix} \cdot (y_D - y_{G_3}) + F_{G_3}^{iy} \cdot (x_D - x_{G_3}) - M_3^i
\end{cases} \quad (2)$$

System (2) can be arranged as a linear system (3) by two equations with two unknowns $R_{12}^x \equiv R_B^x$; $R_{12}^y \equiv R_B^y$, with the coefficients, given from system (4).

$$\begin{cases} a_{11} \cdot R_{12}^x + a_{12} \cdot R_{12}^y = a_1 \\ a_{21} \cdot R_{12}^x + a_{22} \cdot R_{12}^y = a_2 \end{cases} \text{ or } \begin{cases} a_{11} \cdot R_B^x + a_{12} \cdot R_B^y = a_1 \\ a_{21} \cdot R_B^x + a_{22} \cdot R_B^y = a_2 \end{cases} \quad (3)$$

$$\begin{cases}
a_{11} = y_C - y_B;\ a_{12} = -(x_C - x_B); \\
a_1 = -M_1 - F_{G_2}^{ix} \cdot (y_C - y_{G_2}) + \\
+ F_{G_2}^{iy} \cdot (x_C - x_{G_2}) - M_2^i \\
a_{21} = y_D - y_B;\ a_{22} = -(x_D - x_B); \\
a_2 = -M_1 - F_{G_2}^{ix} \cdot (y_D - y_{G_2}) + F_{G_2}^{iy} \cdot (x_D - x_{G_2}) - \\
- M_2^i - M_2 - - F_{G_3}^{ix} \cdot (y_D - y_{G_3}) + \\
+ F_{G_3}^{iy} \cdot (x_D - x_{G_3}) - M_3^i
\end{cases} \quad (4)$$

Solutions of the system (3) will be given by system (5).

$$\begin{cases}
\Delta = \begin{vmatrix} a_{11} & a_{12} \\ a_{21} & a_{22} \end{vmatrix} = a_{11} \cdot a_{22} - a_{12} \cdot a_{21} \quad \Delta_x = \begin{vmatrix} a_1 & a_{12} \\ a_2 & a_{22} \end{vmatrix} = a_{22} \cdot a_1 - a_{12} \cdot a_2 \\
\Delta_y = \begin{vmatrix} a_{11} & a_1 \\ a_{21} & a_2 \end{vmatrix} = a_{11} \cdot a_2 - a_{21} \cdot a_1 \\
R_B^x \equiv R_{12}^x = \dfrac{\Delta_x}{\Delta} = \dfrac{a_{22} \cdot a_1 - a_{12} \cdot a_2}{a_{11} \cdot a_{22} - a_{12} \cdot a_{21}}; \\
R_B^y \equiv R_{12}^y = \dfrac{\Delta_y}{\Delta} = \dfrac{a_{11} \cdot a_2 - a_{21} \cdot a_1}{a_{11} \cdot a_{22} - a_{12} \cdot a_{21}}
\end{cases} \quad (5)$$

Further determine other two internal forces, R_{03}^x și R_{03}^y, or (R_D^x și R_D^y).

Next we write the sum of all forces on the dyad (2,3) designed separately, first on the x axis and then on the y axis, (see the system 6).

$$\begin{cases} \sum F_x^{(2,3)} = 0 \Rightarrow \\ \Rightarrow R_{12}^x + F_{G_2}^{ix} + F_{G_3}^{ix} + R_{03}^x = 0 \Rightarrow \\ \Rightarrow R_D^x \equiv R_{03}^x = -R_{12}^x - F_{G_2}^{ix} - F_{G_3}^{ix} \\ \sum F_y^{(2,3)} = 0 \Rightarrow \\ \Rightarrow R_{12}^y + F_{G_2}^{iy} + F_{G_3}^{iy} + R_{03}^y = 0 \Rightarrow \\ \Rightarrow R_D^y \equiv R_{03}^y = -R_{12}^y - F_{G_2}^{iy} - F_{G_3}^{iy} \end{cases} \quad (6)$$

$$\begin{cases} \begin{cases} \sum F_x^{(2)} = 0 \Rightarrow R_{12}^x + F_{G_2}^{ix} - \\ - R_{23}^x = 0 \Rightarrow R_{23}^x = R_{12}^x + F_{G_2}^{ix} \\ \sum F_y^{(2)} = 0 \Rightarrow R_{12}^y + F_{G_2}^{iy} - \\ - R_{23}^y = 0 \Rightarrow R_{23}^y = R_{12}^y + F_{G_2}^{iy} \end{cases} \\ or \begin{cases} \sum F_x^{(3)} = 0 \Rightarrow R_{23}^x + F_{G_3}^{ix} + \\ + R_D^x = 0 \Rightarrow R_{23}^x = -F_{G_3}^{ix} - R_D^x \\ \sum F_y^{(3)} = 0 \Rightarrow R_{23}^y + F_{G_3}^{iy} + \\ + R_D^y = 0 \Rightarrow R_{23}^y = -F_{G_3}^{iy} - R_D^y \end{cases} \end{cases} \quad (7)$$

For the last two scalar components of the internal force from the joint C, one writes a new balance of forces on element 2 (for example), designed separately on axes x and y (system 7).

We obtained directly the internal forces R_{23}^x and R_{23}^y. Their opposites, R_{32}^x and R_{32}^y, they will be equal but opposite directed their, or in other words will have the same value but opposite sign [3].

For that all kinetostatic calculations of the 3R dyad to be possible, must be determined in advance, the forces and moments of inertia, separately for each element of the dyad. These are called „the group of the inertial forces", and are expressed with the relations system (8).

$$\begin{cases} \begin{cases} F_{G_2}^{ix} = -m_2 \cdot \ddot{x}_{G_2} \\ F_{G_2}^{iy} = -m_2 \cdot \ddot{y}_{G_2} \\ M_2^i = -J_{G_2} \cdot \varepsilon_2 \end{cases} \begin{cases} F_{G_3}^{ix} = -m_3 \cdot \ddot{x}_{G_3} \\ F_{G_3}^{iy} = -m_3 \cdot \ddot{y}_{G_3} \\ M_3^i = -J_{G_3} \cdot \varepsilon_3 \end{cases} \\ \\ \begin{cases} x_{G_2} = x_B + s_2 \cdot \cos\varphi_2 \\ y_{G_2} = y_B + s_2 \cdot \sin\varphi_2 \end{cases} \Rightarrow \\ \Rightarrow \begin{cases} \dot{x}_{G_2} = \dot{x}_B - s_2 \cdot \sin\varphi_2 \cdot \dot{\varphi}_2 \\ \dot{y}_{G_2} = \dot{y}_B + s_2 \cdot \cos\varphi_2 \cdot \dot{\varphi}_2 \end{cases} \Rightarrow \\ \Rightarrow \begin{cases} \ddot{x}_{G_2} = \ddot{x}_B - s_2 \cdot \cos\varphi_2 \cdot \omega_2^2 - s_2 \cdot \sin\varphi_2 \cdot \varepsilon_2 \\ \ddot{y}_{G_2} = \ddot{y}_B - s_2 \cdot \sin\varphi_2 \cdot \omega_2^2 + s_2 \cdot \cos\varphi_2 \cdot \varepsilon_2 \end{cases} \\ \\ \begin{cases} x_{G_3} = x_D + s_{3'} \cdot \cos\varphi_{3'} \\ y_{G_3} = y_D + s_{3'} \cdot \sin\varphi_{3'} \end{cases} \Rightarrow \\ \Rightarrow \begin{cases} \dot{x}_{G_3} = \dot{x}_D - s_{3'} \cdot \sin\varphi_{3'} \cdot \dot{\varphi}_3 \\ \dot{y}_{G_3} = \dot{y}_D + s_{3'} \cdot \cos\varphi_{3'} \cdot \dot{\varphi}_3 \end{cases} \Rightarrow \\ \Rightarrow \begin{cases} \ddot{x}_{G_3} = \ddot{x}_D - s_{3'} \cdot \cos\varphi_{3'} \cdot \omega_3^2 - s_{3'} \cdot \sin\varphi_{3'} \cdot \varepsilon_3 \\ \ddot{y}_{G_3} = \ddot{y}_D - s_{3'} \cdot \sin\varphi_{3'} \cdot \omega_3^2 + s_{3'} \cdot \cos\varphi_{3'} \cdot \varepsilon_3 \end{cases} \end{cases} \quad (8)$$

3. DIAGRAMS OF THE FORCES FROM JOINTS

The joints forces can be determined and represented by the two diagrams below (Figure 2, and 3).

Below you can see the six forces (internal forces) of joints from dyad 3R, depending on the angle of the crank FI, when the dyad is linked together with a crank, forming a mechanism 4R [1-4].

Variation is represented on an entire cycle kinematic, for an angular velocity of crank, 200 or 300 [s^{-1}].

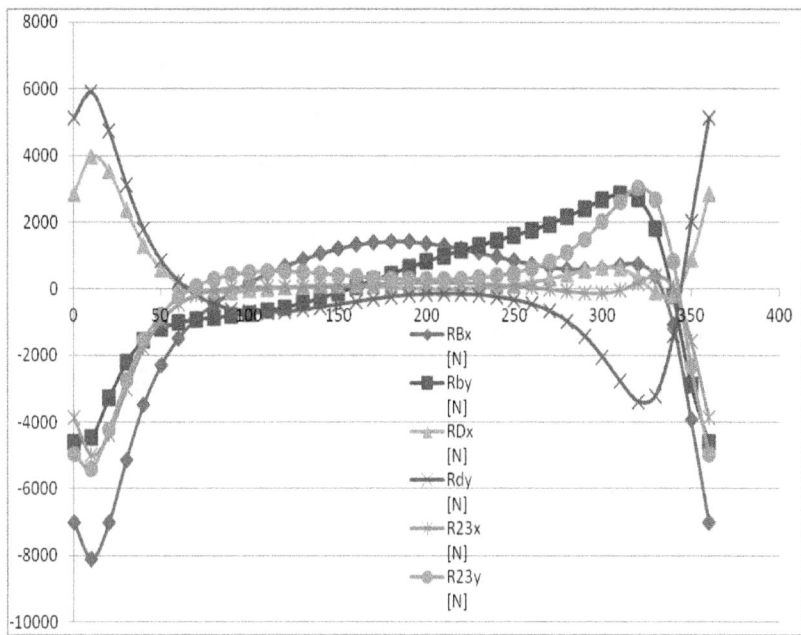

Fig. 2. *The six internal forces of joints;* $\omega = 200 \ [s^{-1}]$

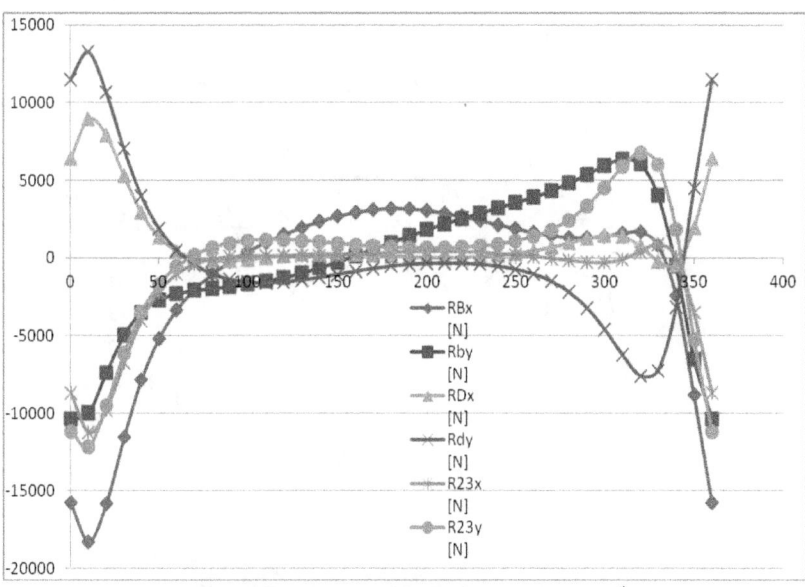

Fig. 3. *The six internal forces of joints;* $\omega = 300 \ [s^{-1}]$

4. CONCLUSIONS

This method presented in the article, is the most elegant and direct method to determine the internal forces at a 3R dyad [3] [1-11].

The method has a strong teaching character.

The relationships presented in this paper allow and the synthesis of robots (the mechanical systems, serial, in movement) [3].

5. IMPORTANCE AND USES

I-The first use of the reaction forces from couplings, is sizing of the kinematic couplings.

II-At the mechanisms with a degree of mobility, with the forces from driving coupling (R_B^x, R_B^y), it determines the required motor torque (M_m). We illustrate by the mechanism articulated quadrilateral (Fig. 4 and relationships 9).

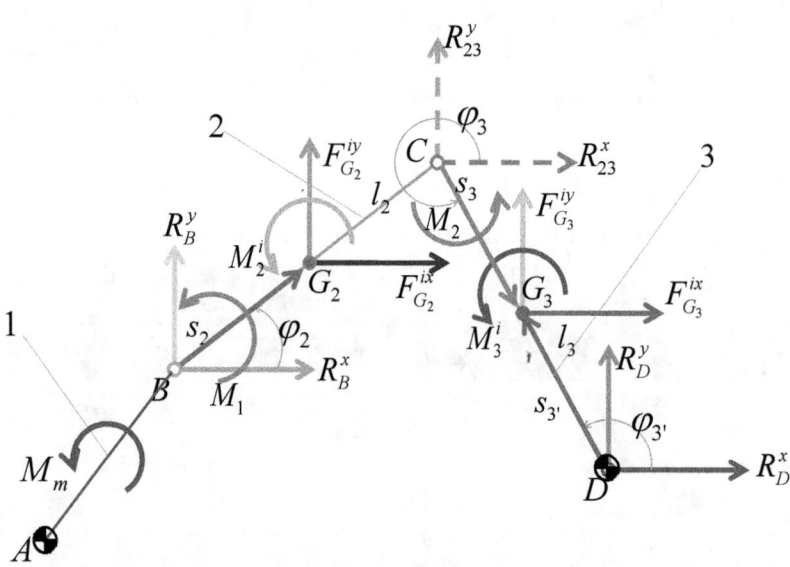

Fig. 4. *The forces at a mechanism articulated quadrilateral*

$$\begin{cases} M_m - R_{21}^x \cdot (y_B - y_A) + R_{21}^y \cdot (x_B - x_A) = 0 \Rightarrow \\ \Rightarrow M_m = R_{21}^x \cdot (y_B - y_A) - R_{21}^y \cdot (x_B - x_A) \Rightarrow \\ \Rightarrow M_m = -R_{12}^x \cdot (y_B - y_A) + R_{12}^y \cdot (x_B - x_A) \Rightarrow \\ \Rightarrow M_m = -R_B^x \cdot (y_B - y_A) + R_B^y \cdot (x_B - x_A) \end{cases} \quad (9)$$

Usually the torques M_1 and M_2 are null. But they can be and an external torque.

III-At the mechanisms with two degree of mobility, with the forces from driving coupling (see the Fig. 5), it determines the required motor torques: $M_1 \equiv M_{m2}, M_2 \equiv M_{m3}$.

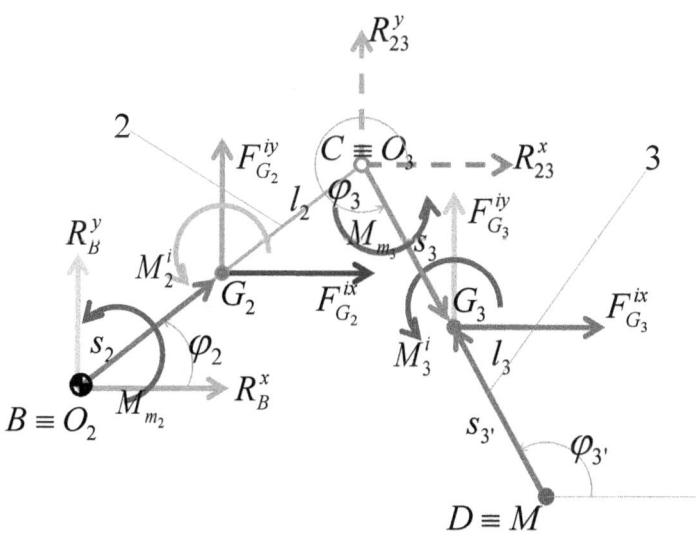

Fig. 5. *The forces at a mechanism with two degree of mobility*

This scheme is used in anthropomorphic robots. Coupling B is denoted by O_2. Coupling C is denoted by O_3. Coupling D become an end effector point M. Basic structure 3R of anthropomorphic robot (Fig. 6) can be decomposed into 2R planar structure (Fig. 5) which also possesses an additional rotating around a vertical axis (O_0O_1).

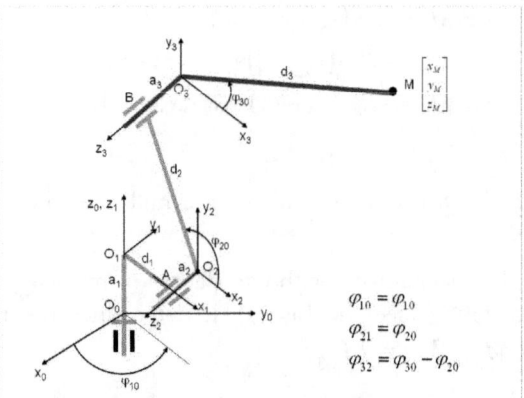

Fig. 6. *The basic structure 3R*

It is more convenient to study the structure plan O_2O_3M system (elements 2 and 3). But since this system (plan, 2R) using balanced, it's good to study in its balanced form (Fig. 7).

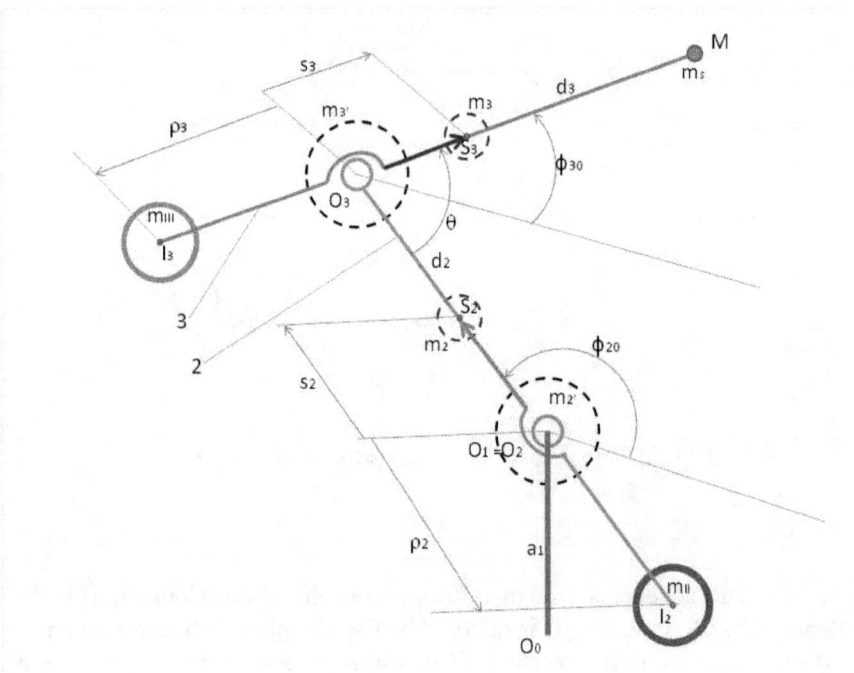

Fig. 7. *The basic, balanced, structure 2R*

Masses and lengths of the system are calculated using the equation 10.

$$\begin{cases} \begin{cases} \sum M_{O_3}^{(3)} = 0 \Rightarrow \\ \Rightarrow m_s \cdot d_3 + m_3 \cdot s_3 = m_{III} \cdot \rho_3 \\ \Rightarrow \rho_3 = \dfrac{m_s \cdot d_3 + m_3 \cdot s_3}{m_{III}} \\ m_{3'} = m_3 + m_s + m_{III} \end{cases} \\ \begin{cases} \sum M_{O_2}^{(2+3)} = 0 \Rightarrow \\ \Rightarrow m_{3'} \cdot d_2 + m_2 \cdot s_2 = m_{II} \cdot \rho_2 \\ \Rightarrow \rho_2 = \dfrac{m_{3'} \cdot d_2 + m_2 \cdot s_2}{m_{II}} \\ m_{2'} = m_{3'} + m_2 + m_{II} \end{cases} \end{cases} \qquad (10)$$

Forces from the driveline balanced plan can be seen in the Fig. 8.

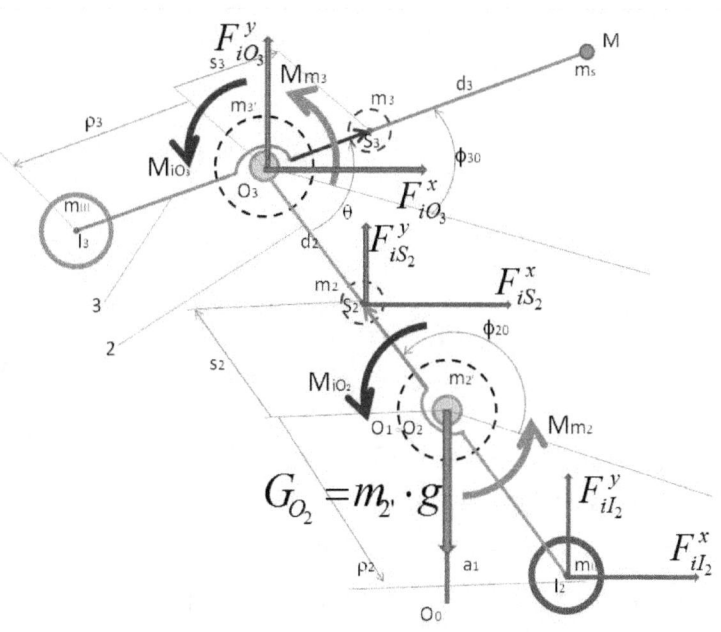

Fig. 8. *The forces of the basic (balanced) structure 2R*

Now, it still writing inertial forces (relations system 11) of the point O_3.

$$\begin{cases} F^x_{iO_3} = -m_{3'} \cdot \ddot{x}_{O_3} = \\ = -m_{3'} \cdot (-)d_2 \cdot \cos\varphi_{20} \cdot \omega^2_{20} = \\ = m_{3'} \cdot d_2 \cdot \cos\varphi_{20} \cdot \omega^2_{20} \\ \\ F^y_{iO_3} = -m_{3'} \cdot \ddot{y}_{O_3} = \\ = -m_{3'} \cdot (-)d_2 \cdot \sin\varphi_{20} \cdot \omega^2_{20} = \\ = m_{3'} \cdot d_2 \cdot \sin\varphi_{20} \cdot \omega^2_{20} \\ \\ M_{iO_3} = -J_{O_3} \cdot \varepsilon_3 \end{cases} \quad (11)$$

Now we are writing and the inertial forces of the points S_2 (12) and I_2 (13).

$$\begin{cases} F^x_{iS_2} = -m_2 \cdot \ddot{x}_{S_2} = m_2 \cdot s_2 \cdot \cos\varphi_{20} \cdot \omega^2_{20} \\ F^y_{iS_2} = -m_2 \cdot \ddot{y}_{S_2} = m_2 \cdot s_2 \cdot \sin\varphi_{20} \cdot \omega^2_{20} \end{cases} \quad (12)$$

$$\begin{cases} F^x_{iI_2} = -m_{II} \cdot \ddot{x}_{I_2} = -m_{II} \cdot \rho_2 \cdot \cos\varphi_{20} \cdot \omega^2_{20} \\ F^y_{iI_2} = -m_{II} \cdot \ddot{y}_{I_2} = -m_{II} \cdot \rho_2 \cdot \sin\varphi_{20} \cdot \omega^2_{20} \end{cases} \quad (13)$$

Now we can write the equilibrium equations on the element 2 projected on the x (system 14) and y (system 15).

$$\begin{cases} \sum F^x_{(2)} = 0 \Rightarrow m_{3'} \cdot d_2 \cdot \cos\varphi_{20} \cdot \omega^2_{20} + m_2 \cdot s_2 \cdot \cos\varphi_{20} \cdot \omega^2_{20} - \\ - m_{II} \cdot \rho_2 \cdot \cos\varphi_{20} \cdot \omega^2_{20} + R^x_{O_2} = 0 \Rightarrow \\ \Rightarrow (m_{3'} \cdot d_2 + m_2 \cdot s_2 - m_{II} \cdot \rho_{II}) \cdot \cos\varphi_{20} \cdot \omega^2_{20} + R^x_{12} = 0 \\ \text{but } m_{3'} \cdot d_2 + m_2 \cdot s_2 - m_{II} \cdot \rho_{II} = 0 \text{ because balanced} \Rightarrow \\ \Rightarrow R^x_{O_2} \equiv R^x_{12} = 0 \end{cases} \quad (14)$$

$$\begin{cases} \sum F_{(2)}^y = 0 \Rightarrow m_{3'} \cdot d_2 \cdot \sin\varphi_{20} \cdot \omega_{20}^2 + m_2 \cdot s_2 \cdot \sin\varphi_{20} \cdot \omega_{20}^2 - \\ - m_{II} \cdot \rho_2 \cdot \sin\varphi_{20} \cdot \omega_{20}^2 - m_{2'} \cdot g + R_{12}^y = 0 \Rightarrow \\ \Rightarrow (m_{3'} \cdot d_2 + m_2 \cdot s_2 - m_{II} \cdot \rho_{II}) \cdot \sin\varphi_{20} \cdot \omega_{20}^2 - m_{2'} \cdot g + R_{12}^y = 0 \\ \text{but } m_{3'} \cdot d_2 + m_2 \cdot s_2 - m_{II} \cdot \rho_{II} = 0 \text{ because balanced} \Rightarrow \\ \Rightarrow R_{O_2}^y \equiv R_{12}^y = m_{2'} \cdot g = G_{O_2} \end{cases} \quad (15)$$

It can be seen that the torque loads are minimal precisely because balancing. Effect given inertial forces (torques produced by these forces) cancel (balance due).

Torques produced by the forces of gravity is canceled and they all balance due.

Balanced final weight also makes the powertrain only one effect, a vertical load (causes a vertical reactor) in fixed coupling.

At a total balanced, even the horizontal load disappears.

It will still write an amount of moments to the fixed point O_2, on the element 2 (system 16).

$$\begin{cases} \sum M_{O_2}^{(2)} = 0 \Rightarrow M_{m_2} - F_{iO_3}^x \cdot d_2 \cdot \cos\left(\varphi_{20} - \dfrac{\pi}{2}\right) - \\ - F_{iO_3}^y \cdot d_2 \cdot \sin\left(\varphi_{20} - \dfrac{\pi}{2}\right) - F_{iS_2}^x \cdot s_2 \cdot \sin\varphi_{20} - F_{iS_2}^y \cdot s_2 \cdot -\cos\varphi_{20} + \\ + F_{iI_2}^x \cdot \rho_2 \cdot \cos\left(\varphi_{20} - \dfrac{\pi}{2}\right) + F_{iI_2}^y \cdot \rho_2 \cdot \sin\left(\varphi_{20} - \dfrac{\pi}{2}\right) + M_{iO_2} = 0 \Rightarrow \\ \Rightarrow M_{m_2} - m_{3'} d_2^2 \omega_{20}^2 \cos\varphi_{20} \sin\varphi_{20} + m_{3'} \cdot d_2^2 \omega_{20}^2 \sin\varphi_{20} \cos\varphi_{20} - \\ - m_2 \cdot s_2^2 \cdot \omega_{20}^2 \cdot \cos\varphi_{20} \cdot \sin\varphi_{20} + m_2 \cdot s_2^2 \cdot \omega_{20}^2 \cdot \sin\varphi_{20} \cdot \cos\varphi_{20} - \\ - m_{II} \cdot \rho_2^2 \cdot \omega_{20}^2 \cos\varphi_{20} \cdot \sin\varphi_{20} + m_{II} \cdot \rho_2^2 \cdot \omega_{20}^2 \cdot \sin\varphi_{20} \cdot \cos\varphi_{20} - \\ - J_{O_2}^* \cdot \varepsilon_2 = 0 \Rightarrow M_{m_2} - J_{O_2}^* \cdot \varepsilon_2 = 0 \Rightarrow M_{m_2} = J_{O_2}^* \cdot \varepsilon_2 \end{cases} \quad (16)$$

Mass moment of inertia (or mechanical) of the element 2, is calculated with relation 17.

$$J_{O_2}^* = J_{O_2} + m_{3'} \cdot d_2^2 = m_2 \cdot s_2^2 + m_{II} \cdot \rho_2^2 + m_{3'} \cdot d_2^2 \quad (17)$$

One can determine now the torque required (M_{m2}), which must be generated by the actuator 2 (mounted in coupling O_2); see the relation (18).

$$M_{m_2} = J^*_{O_2} \cdot \varepsilon_2 = \left(m_2 \cdot s_2^2 + m_{II} \cdot \rho_2^2 + m_{3'} \cdot d_2^2\right) \cdot \ddot{\varphi}_{20} \qquad (18)$$

We now sum of the moments of all forces on item 3 in relation to swivel O_3 (relationship 19).

$$\sum M^{(3)}_{O_3} = 0 \Rightarrow$$
$$M_{m_3} + M_{iO_3} = 0 \Rightarrow M_{m_3} - J_{O_3} \cdot \varepsilon_3 = 0 \Rightarrow$$
$$\Rightarrow M_{m_3} = J_{O_3} \cdot \varepsilon_3 \Rightarrow \qquad (19)$$
$$\Rightarrow M_{m_3} = \left(m_s \cdot d_3^2 + m_3 \cdot s_3^2 + m_{III} \cdot \rho_3^2\right) \cdot \ddot{\varphi}_{30}$$

One determines now and the vertical component, of the reaction, from the mobile (internal) coupling O_3; (see the relations of the system 20).

$$\begin{cases} \sum F^y_{(3)} = 0 \Rightarrow -m_{3'} \cdot g + R^y_{23} = 0 \Rightarrow \\ \Rightarrow R^y_{23} = m_{3'} \cdot g \Rightarrow \\ \Rightarrow R^y_{32} = -R^y_{23} = -m_{3'} \cdot g \end{cases} \qquad (20)$$

Horizontal component (of the reaction from the kinematic coupling O_3) is zero (21).

$$R^x_{23} = -R^y_{32} = 0 \qquad (21)$$

6. DYNAMICS OF SYSTEM 2R (LAGRANGE DIFFERENTIAL EQUATION OF THE SECOND KIND)

It writes now, just the most important relations of the system 2R, in the form 22.

$$\begin{cases} M_{m_2} = J^*_{O_2} \cdot \varepsilon_2 \\ M_{m_3} = J_{O_3} \cdot \varepsilon_3 \\ \\ M_{m_2} = \left(m_2 \cdot s_2^2 + m_{II} \cdot \rho_2^2 + m_{3'} \cdot d_2^2\right) \cdot \ddot{\varphi}_{20} \\ M_{m_3} = \left(m_s \cdot d_3^2 + m_3 \cdot s_3^2 + m_{III} \cdot \rho_3^2\right) \cdot \ddot{\varphi}_{30} \end{cases} \qquad (22)$$

These relationships necessary to study the dynamics of the kinematic chain level (22), can be obtained directly by another method, which uses Lagrange differential equation of the second kind, and the kinetic energy saving mechanism.

This method is more direct than cinetostatic study, but has the disadvantage of not determining the loadings (reactions, internal forces) from kinematics chain, necessary to calculate the strength of the material in applications in which certain dimensions are selected (thickness or diameter) of the kinematic elements 2 and 3, and connecting joints.

One first determines the speeds, in the gravity centers (relations from system 23, and Fig. 9).

$$\begin{cases} \dot{x}_{O_2} = 0; \\ \\ \dot{y}_{O_2} = 0; \quad \dot{\varphi}_{20} \equiv \omega_{20} \equiv \omega_2 \\ \\ \dot{x}_{O_3} = -d_2 \cdot \sin \varphi_{20} \cdot \omega_2; \\ \dot{y}_{O_3} = d_2 \cdot \cos \varphi_{20} \cdot \omega_2; \\ \\ \dot{\varphi}_{30} \equiv \omega_{30} \equiv \omega_3 \end{cases} \qquad (23)$$

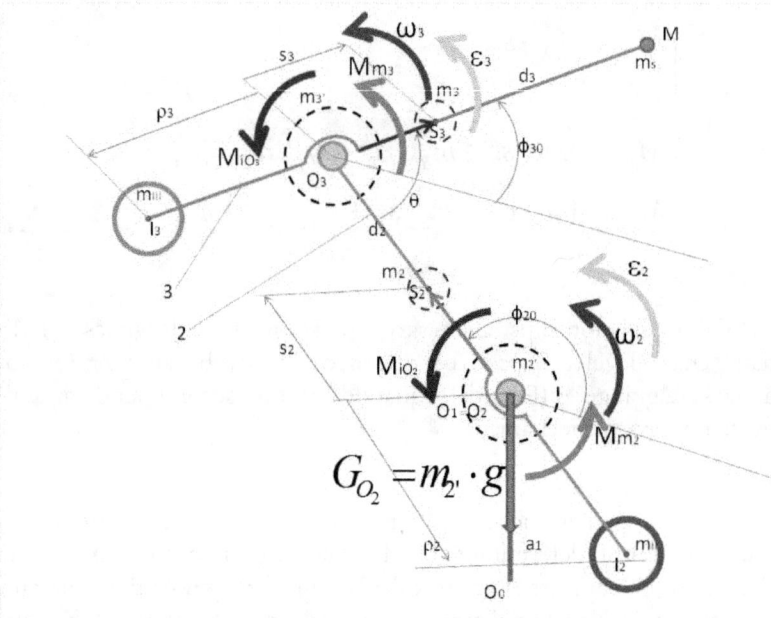

Fig. 9. *Dynamics of the driveline balanced plan*

For item 3, mass moment of inertia or mechanical (inertial mass) is determined by the relationship 24.

$$J_{O_3} = m_s \cdot d_3^2 + m_3 \cdot s_3^2 + m_{III} \cdot \rho_3^2 \tag{24}$$

For item 2, will cause mass moment of inertia (mechanical) in fixed joint O_2 (25).

$$J_{O_2} = m_2 \cdot s_2^2 + m_{II} \cdot \rho_2^2 \tag{25}$$

The kinetic energy of mechanism is determined with the relations of the system (26).

$$\begin{cases} E = \frac{1}{2} \cdot J_{O_2} \cdot \omega_2^2 + \frac{1}{2} \cdot J_{O_3} \cdot \omega_3^2 + \\ + \frac{1}{2} \cdot m_{3'} \cdot \dot{x}_{O_3}^2 + \frac{1}{2} \cdot m_{3'} \cdot \dot{y}_{O_3}^2 = \\ = \frac{1}{2} \cdot J_{O_2} \cdot \omega_2^2 + \frac{1}{2} \cdot J_{O_3} \cdot \omega_3^2 + \\ + \frac{1}{2} \cdot m_{3'} \cdot d_2^2 \cdot \omega_2^2 = \frac{1}{2} \cdot J_{O_3} \cdot \omega_3^2 + \\ + \frac{1}{2} \cdot \omega_2^2 \cdot \left(J_{O_2} + m_{3'} \cdot d_2^2 \right) = \\ = \frac{1}{2} \cdot J_{O_3} \cdot \omega_3^2 + \frac{1}{2} \cdot J_{O_2}^* \cdot \omega_2^2 \\ J_{O_2}^* = J_{O_2} + m_{3'} \cdot d_2^2 \end{cases} \quad (26)$$

Kinetic energy equation for the balanced driveline is expressed with final relationship (27).

$$E = \frac{1}{2} \cdot J_{O_3} \cdot \omega_3^2 + \frac{1}{2} \cdot J_{O_2}^* \cdot \omega_2^2 \quad (27)$$

It uses the Lagrange differential equations of second kind (28).

$$\begin{cases} \frac{d}{dt}\left(\frac{\partial E}{\partial \dot{q}_k} \right) - \frac{\partial E}{\partial q_k} = Q_k \\ \text{with } k = 2,\ 3 \\ \\ \frac{d}{dt}\left(\frac{\partial E}{\partial \dot{q}_2} \right) - \frac{\partial E}{\partial q_2} = Q_2 \\ \frac{d}{dt}\left(\frac{\partial E}{\partial \dot{q}_3} \right) - \frac{\partial E}{\partial q_3} = Q_3 \end{cases} \quad (28)$$

How kinetic energy in this case does not depend directly from the kinematic parameters of positions q₂ and q₃, represented by the position angles φ₂₀ and φ₃₀, it can be used the simplified form of the Lagrange equations (29).

$$\begin{cases} \dfrac{d}{dt}\left(\dfrac{\partial E}{\partial \dot{q}_k}\right) = Q_k \quad \text{with} \quad k = 2, 3 \\[6pt] \dfrac{d}{dt}\left(\dfrac{\partial E}{\partial \dot{q}_2}\right) = Q_2 \Rightarrow \dfrac{d}{dt}\left(\dfrac{\partial E}{\partial \omega_2}\right) = M_{m_2} \\[6pt] \dfrac{d}{dt}\left(\dfrac{\partial E}{\partial \dot{q}_3}\right) = Q_3 \Rightarrow \dfrac{d}{dt}\left(\dfrac{\partial E}{\partial \omega_3}\right) = M_{m_3} \end{cases} \qquad (29)$$

By replacing the partial derivatives and making the derivatives in function of time, the system (29) takes the form (30).

$$\begin{cases} \dfrac{\partial E}{\partial \omega_2} = J^*_{O_2} \cdot \omega_2 \Rightarrow \dfrac{d}{dt}\left(\dfrac{\partial E}{\partial \omega_2}\right) = J^*_{O_2} \cdot \varepsilon_2 \Rightarrow J^*_{O_2} \cdot \varepsilon_2 = M_{m_2} \\[6pt] \dfrac{\partial E}{\partial \omega_3} = J_{O_3} \cdot \omega_3 \Rightarrow \dfrac{d}{dt}\left(\dfrac{\partial E}{\partial \omega_3}\right) = J_{O_3} \cdot \varepsilon_3 \Rightarrow J_{O_3} \cdot \varepsilon_3 = M_{m_3} \\[6pt] J^*_{O_2} \cdot \varepsilon_2 = M_{m_2} \\[6pt] J_{O_3} \cdot \varepsilon_3 = M_{m_3} \\[6pt] J^*_{O_2} = m_2 \cdot s_2^2 + m_{II} \cdot \rho_2^2 + m_{3'} \cdot d_2^2 \\[6pt] J_{O_3} = m_s \cdot d_3^2 + m_3 \cdot s_3^2 + m_{III} \cdot \rho_3^2 \\[6pt] M_{m_2} = \left(m_2 \cdot s_2^2 + m_{II} \cdot \rho_2^2 + m_{3'} \cdot d_2^2\right) \cdot \varepsilon_2 \\[6pt] M_{m_3} = \left(m_s \cdot d_3^2 + m_3 \cdot s_3^2 + m_{III} \cdot \rho_3^2\right) \cdot \varepsilon_3 \end{cases} \qquad (30)$$

REFERENCES

[1] Comanescu, A., ş.a., *Bazele modelarii mecanismelor.* E.Politeh. Press, Bucureşti, 2010, ISBN 13: 978-606-515-114-7, page count 274.
[2] Franklin, D.J., *Ingenious Mechanisms for Designers and Inventors*, Industrial Press publisher, 1930, ISBN-13: 978-0831110840.
[3] Petrescu, F.I., *Teoria Mecanismelor si a Masinilor*, Create Space Publisher, USA, 2011, ISBN/EAN 13: 1468015826 / 978-1-4680-1582-9, page count 432.
[4] Waldron, K.J., Kinzel, G.L., *Kinematics, Dynamics, and Design of Machinery*, Department of Mechanical Engineeering, Ohio State University, USA, Available in:
http://www.me.mtu.edu/~nels/3501SUMMER07/kinematic_matlab_help.pdf
[5] Petrescu F.I., Petrescu R.V., *Kinematics of the Planar Quadrilateral Mechanism*, Engevista Journal, Vol. 14, N. 3, December 2012, p. 345-348. Available in: http://www.uff.br/engevista/seer/index.php/engevista/article/viewArticle/377
[6] Powell I.L., B.A.Miere, *The kinematic analysis and simulation of the parallel topology manipulator*, The Marconi Review, 1982.
[7] Reddy M., a.o., *Precise Non Linear Modeling of Flexible Link Flexible Joint Manipulator*, IREMOS, Vol. 5, N. 3, June (Part B) 2012, p. 1368-1374.
[8] Shanmuga G., a.o., *A Survey on Development of Inspection Robots: Kinematic Analysis, Workspace Simulation and Software Development*, IREME, Vol. 6, N. 7, November 2012, p.1493-1507.
[9] Petrescu, F.I., Petrescu, R.V., *Forces and Efficiency of Cams*, International Review of Mechanical Engineering (IREME Journal), March 2013, Vol. 7, N. 3, ISSN: 1970-8734, p., 2013 (Indexed SCOPUS).
[10] Prazeres U.R., Santos W.L.R., *Caracterizacao mecanica e electrica da liga 6101 modificada com diferentes teores de MG*, Engevista Journal, Vol. 13, N. 2, 2011, p. 122-128. Available in:
http://www.uff.br/engevista/seer/index.php/engevista/article/view/260
[11] Nogueira O.C., Real M.V., *Estudo Comparativo de Motores Diesel Maritimos Atraves da Analise de Lubrificantes Usados e Engenharia de Confiabilidade*, Engevista Journal, Vol. 13, N. 3, December 2011, p. 244-254. Available in:
http://www.uff.br/engevista/seer/index.php/engevista/article/view/300

Chapter 08_Parallel Moving Mechanical Systems

1. INTRODUCTION

Moving mechanical structures are used increasingly in almost all vital sectors of humanity (Cao, 2013). The robots are able to process integrated circuits sizes micro and nano, on which the man they can be seen even with electron microscopy (Garcia, 2007). Dyeing parts in toxic environments (Tong, 2013), working in chemical and radioactive environments, or at depths and pressures at the bottom of huge oceans, or even cosmic space conquest and visiting exo-planets, are now possible, and were turned into from the dream in reality, because mechanical platforms sequential gearbox (Perumaal, 2013).

Robots were developed and diversified, different aspects, but today, they start to be directed on two major categories: systems serial and parallel systems (Padula, 2013). Parallel systems are more solid, but more difficult to designed and handled, which serial systems were those which have developed the most. In medical operations or radioactive environments are preferred mobile systems parallel to their high accuracy positioning (Reddy, 2012).

2. THE STRUCTURE AND GEOMETRY OF A STEWARD SYSTEM

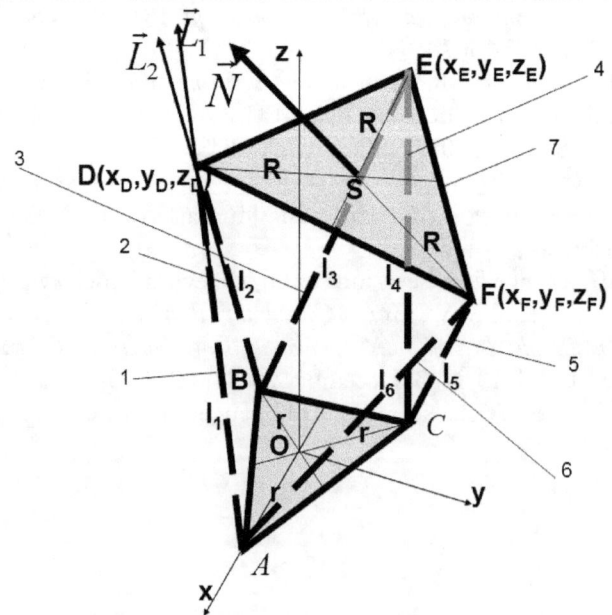

Figure 1. *The basic structure of a Stewart System*

Figure 1 shows unit vectors route along items 1 and 2 from the bottom to mobile platform. The co-ordinates of vectors unit belonging to motto-items 1-6 (variable-length) are given by the system (1).

$$\begin{cases} \alpha_1 = \dfrac{x_D - x_A}{l_1}; & \beta_1 = \dfrac{y_D - y_A}{l_1}; & \gamma_1 = \dfrac{z_D - z_A}{l_1}; \\[6pt] \alpha_2 = \dfrac{x_D - x_B}{l_2}; & \beta_2 = \dfrac{y_D - y_B}{l_2}; & \gamma_2 = \dfrac{z_D - z_B}{l_2}; \\[6pt] \alpha_3 = \dfrac{x_E - x_B}{l_3}; & \beta_3 = \dfrac{y_E - y_B}{l_3}; & \gamma_3 = \dfrac{z_E - z_B}{l_3}; \\[6pt] \alpha_4 = \dfrac{x_E - x_C}{l_4}; & \beta_4 = \dfrac{y_E - y_C}{l_4}; & \gamma_4 = \dfrac{z_E - z_C}{l_4}; \\[6pt] \alpha_5 = \dfrac{x_F - x_C}{l_5}; & \beta_5 = \dfrac{y_F - y_C}{l_5}; & \gamma_5 = \dfrac{z_F - z_C}{l_5}; \\[6pt] \alpha_6 = \dfrac{x_F - x_A}{l_6}; & \beta_6 = \dfrac{y_F - y_A}{l_6}; & \gamma_6 = \dfrac{z_F - z_A}{l_6}; \end{cases} \quad (1)$$

Where these lengths of vectors unit are given by the system (2), and actual lengths of the six mottoelements (variables) is expressed by the system (3).

$$\begin{cases} \bar{L}_1 = \alpha_1 \cdot \bar{i} + \beta_1 \cdot \bar{j} + \gamma_1 \cdot \bar{k}; & \bar{L}_2 = \alpha_2 \cdot \bar{i} + \beta_2 \cdot \bar{j} + \gamma_2 \cdot \bar{k}; \\ \bar{L}_3 = \alpha_3 \cdot \bar{i} + \beta_3 \cdot \bar{j} + \gamma_3 \cdot \bar{k}; & \bar{L}_4 = \alpha_4 \cdot \bar{i} + \beta_4 \cdot \bar{j} + \gamma_4 \cdot \bar{k}; \\ \bar{L}_5 = \alpha_5 \cdot \bar{i} + \beta_5 \cdot \bar{j} + \gamma_5 \cdot \bar{k}; & \bar{L}_6 = \alpha_6 \cdot \bar{i} + \beta_6 \cdot \bar{j} + \gamma_6 \cdot \bar{k} \end{cases} \quad (2)$$

$$\begin{cases} \bar{l}_1 = l_1 \cdot \bar{L}_1 = \alpha_1 \cdot l_1 \cdot \bar{i} + \beta_1 \cdot l_1 \cdot \bar{j} + \gamma_1 \cdot l_1 \cdot \bar{k}; \\ \bar{l}_2 = l_2 \cdot \bar{L}_2 = \alpha_2 \cdot l_2 \cdot \bar{i} + \beta_2 \cdot l_2 \cdot \bar{j} + \gamma_2 \cdot l_2 \cdot \bar{k}; \\ \bar{l}_3 = l_3 \cdot \bar{L}_3 = \alpha_3 \cdot l_3 \cdot \bar{i} + \beta_3 \cdot l_3 \cdot \bar{j} + \gamma_3 \cdot l_3 \cdot \bar{k}; \\ \bar{l}_4 = l_4 \cdot \bar{L}_4 = \alpha_4 \cdot l_4 \cdot \bar{i} + \beta_4 \cdot l_4 \cdot \bar{j} + \gamma_4 \cdot l_4 \cdot \bar{k}; \\ \bar{l}_5 = l_5 \cdot \bar{L}_5 = \alpha_5 \cdot l_5 \cdot \bar{i} + \beta_5 \cdot l_5 \cdot \bar{j} + \gamma_5 \cdot l_5 \cdot \bar{k}; \\ \bar{l}_6 = l_6 \cdot \bar{L}_6 = \alpha_6 \cdot l_6 \cdot \bar{i} + \beta_6 \cdot l_6 \cdot \bar{j} + \gamma_6 \cdot l_6 \cdot \bar{k} \end{cases} \quad (3)$$

In Figure 2 is represented a motto element (motto element 1) in a position snapshots. If a structural mottoelement consists of two moving elements which translates relative, drive train and especially dynamic it is more convenient to represent the mottoelement as a single moving components. We have thus seven moving parts (the six motoelements or feet to which is added mobile platform 7) and one fixed.

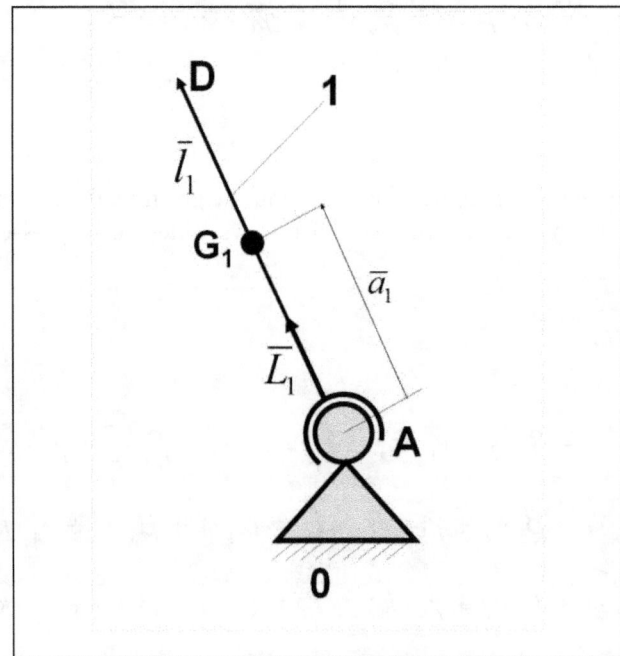

Figure 2. *The basic structure of a motto element*

For the stem 1, one writes relations (4-7). The length l1 is variable; in the same way and the distance a1 which defines the position of the center point of gravity G1 (and the center of gravity G1 is continuously changed, even if rod mass formed from virtually two kinematic elements in relative movement of translation is virtually constant).

$$\begin{cases} \alpha_1 \cdot l_1 = x_D - x_A; & \dot{\alpha}_1 \cdot l_1 + \alpha_1 \cdot \dot{l}_1 = \dot{x}_D; & \dot{\alpha}_1 = \dfrac{\dot{x}_D - \alpha_1 \cdot \dot{l}_1}{l_1}; \\ \beta_1 \cdot l_1 = y_D - y_A; & \dot{\beta}_1 \cdot l_1 + \beta_1 \cdot \dot{l}_1 = \dot{y}_D; & \dot{\beta}_1 = \dfrac{\dot{y}_D - \beta_1 \cdot \dot{l}_1}{l_1}; \\ \gamma_1 \cdot l_1 = z_D - z_A; & \dot{\gamma}_1 \cdot l_1 + \gamma_1 \cdot \dot{l}_1 = \dot{z}_D; & \dot{\gamma}_1 = \dfrac{\dot{z}_D - \gamma_1 \cdot \dot{l}_1}{l_1} \end{cases} \quad (4)$$

$$\begin{cases} x_D = x_A + \alpha_1 \cdot l_1; & y_D = y_A + \beta_1 \cdot l_1; & z_D = z_A + \gamma_1 \cdot l_1; \\ x_{G_1} = x_A + \alpha_1 \cdot a_1; & y_{G_1} = y_A + \beta_1 \cdot a_1; & z_{G_1} = z_A + \gamma_1 \cdot a_1 \end{cases} \quad (5)$$

$$\begin{cases} x_{G_1} = \dfrac{a_1 \cdot x_D + (l_1 - a_1) \cdot x_A}{l_1}; \\ y_{G_1} = \dfrac{a_1 \cdot y_D + (l_1 - a_1) \cdot y_A}{l_1}; \\ z_{G_1} = \dfrac{a_1 \cdot z_D + (l_1 - a_1) \cdot z_A}{l_1} \end{cases} \quad (6)$$

$$\begin{cases} l_1 \cdot x_{G_1} = a_1 \cdot x_D + (l_1 - a_1) \cdot x_A; \dot{l}_1 \cdot x_{G_1} + l_1 \cdot \dot{x}_{G_1} = \\ = \dot{a}_1 \cdot x_D + a_1 \cdot \dot{x}_D + (\dot{l}_1 - \dot{a}_1) \cdot x_A; \\ \dot{x}_{G_1} = \dfrac{\dot{a}_1 \cdot x_D + a_1 \cdot \dot{x}_D - \dot{l}_1 \cdot x_{G_1} + (\dot{l}_1 - \dot{a}_1) \cdot x_A}{l_1}; \\ \dot{y}_{G_1} = \dfrac{\dot{a}_1 \cdot y_D + a_1 \cdot \dot{y}_D - \dot{l}_1 \cdot y_{G_1} + (\dot{l}_1 - \dot{a}_1) \cdot y_A}{l_1}; \\ \dot{z}_{G_1} = \dfrac{\dot{a}_1 \cdot z_D + a_1 \cdot \dot{z}_D - \dot{l}_1 \cdot z_{G_1} + (\dot{l}_1 - \dot{a}_1) \cdot z_A}{l_1} \end{cases} \quad (7)$$

Kinetic energy of the mechanism (8) is being written while taking account of the fact that the translation center of gravity of each mottoelement already contains and the effect of different rotations. Each motoelement (rod) will be studied as a single kinematic element variable-length to constant mass and the position of the center of gravity variable. Each mottoelement movement is one of spatial rotation (Petrescu, 2009-2013).

$$\begin{cases} E_c = \dfrac{m_1}{2} \cdot \left(\dot{x}_{G_1}^2 + \dot{y}_{G_1}^2 + \dot{z}_{G_1}^2 \right) + \dfrac{m_2}{2} \cdot \left(\dot{x}_{G_2}^2 + \dot{y}_{G_2}^2 + \dot{z}_{G_2}^2 \right) + \\ + \dfrac{m_3}{2} \cdot \left(\dot{x}_{G_3}^2 + \dot{y}_{G_3}^2 + \dot{z}_{G_3}^2 \right) + \dfrac{m_4}{2} \cdot \left(\dot{x}_{G_4}^2 + \dot{y}_{G_4}^2 + \dot{z}_{G_4}^2 \right) + \\ + \dfrac{m_5}{2} \cdot \left(\dot{x}_{G_5}^2 + \dot{y}_{G_5}^2 + \dot{z}_{G_5}^2 \right) + \dfrac{m_6}{2} \cdot \left(\dot{x}_{G_6}^2 + \dot{y}_{G_6}^2 + \dot{z}_{G_6}^2 \right) + \\ + \dfrac{m_7}{2} \cdot \left(\dot{x}_S^2 + \dot{y}_S^2 + \dot{z}_S^2 \right) + \dfrac{J_{7SN}}{2} \cdot \omega_{7SN}^2 \end{cases} \quad (8)$$

After the model system (7) is determined velocities of centers of the weight of the six rods (see equations 9). Speeds $\dot{x}_S$, $\dot{y}_S$, $\dot{z}_S$, ω_{7SN} are known. The masses are weighed and mass moment of inertia after axis N shall be calculated on the basis of a approximate formula (10).

$$\begin{cases}
\dot{x}_{G_1} = \dfrac{\dot{a}_1 \cdot (x_D - x_A) + a_1 \cdot \dot{x}_D + \dot{l}_1 \cdot (x_A - x_{G_1})}{l_1}; \dot{y}_{G_1} = \dfrac{\dot{a}_1 \cdot (y_D - y_A) + a_1 \cdot \dot{y}_D + \dot{l}_1 \cdot (y_A - y_{G_1})}{l_1}; \\
\dot{z}_{G_1} = \dfrac{\dot{a}_1 \cdot (z_D - z_A) + a_1 \cdot \dot{z}_D + \dot{l}_1 \cdot (z_A - z_{G_1})}{l_1}; \dot{x}_{G_2} = \dfrac{\dot{a}_2 \cdot (x_D - x_B) + a_2 \cdot \dot{x}_D + \dot{l}_2 \cdot (x_B - x_{G_2})}{l_2} \\
\dot{y}_{G_2} = \dfrac{\dot{a}_2 \cdot (y_D - y_B) + a_2 \cdot \dot{y}_D + \dot{l}_2 \cdot (y_B - y_{G_2})}{l_2}; \dot{z}_{G_2} = \dfrac{\dot{a}_2 \cdot (z_D - z_B) + a_2 \cdot \dot{z}_D + \dot{l}_2 \cdot (z_B - z_{G_2})}{l_2}; \\
\dot{x}_{G_3} = \dfrac{\dot{a}_3 \cdot (x_E - x_B) + a_3 \cdot \dot{x}_E + \dot{l}_3 \cdot (x_B - x_{G_3})}{l_3}; \dot{y}_{G_3} = \dfrac{\dot{a}_3 \cdot (y_E - y_B) + a_3 \cdot \dot{y}_E + \dot{l}_3 \cdot (y_B - y_{G_3})}{l_3}; \\
\dot{z}_{G_3} = \dfrac{\dot{a}_3 \cdot (z_E - z_B) + a_3 \cdot \dot{z}_E + \dot{l}_3 \cdot (z_B - z_{G_3})}{l_3}; \dot{x}_{G_4} = \dfrac{\dot{a}_4 \cdot (x_E - x_C) + a_4 \cdot \dot{x}_E + \dot{l}_4 \cdot (x_C - x_{G_4})}{l_4}; \\
\dot{y}_{G_4} = \dfrac{\dot{a}_4 \cdot (y_E - y_C) + a_4 \cdot \dot{y}_E + \dot{l}_4 \cdot (y_C - y_{G_4})}{l_4}; \dot{z}_{G_4} = \dfrac{\dot{a}_4 \cdot (z_E - z_C) + a_4 \cdot \dot{z}_E + \dot{l}_4 \cdot (z_C - z_{G_4})}{l_4}; \\
\dot{x}_{G_5} = \dfrac{\dot{a}_5 \cdot (x_F - x_C) + a_5 \cdot \dot{x}_F + \dot{l}_5 \cdot (x_C - x_{G_5})}{l_5}; \dot{y}_{G_5} = \dfrac{\dot{a}_5 \cdot (y_F - y_C) + a_5 \cdot \dot{y}_F + \dot{l}_5 \cdot (y_C - y_{G_5})}{l_5}; \\
\dot{z}_{G_5} = \dfrac{\dot{a}_5 \cdot (z_F - z_C) + a_5 \cdot \dot{z}_F + \dot{l}_5 \cdot (z_C - z_{G_5})}{l_5}; \dot{x}_{G_6} = \dfrac{\dot{a}_6 \cdot (x_F - x_A) + a_6 \cdot \dot{x}_F + \dot{l}_6 \cdot (x_A - x_{G_6})}{l_6}; \\
\dot{y}_{G_6} = \dfrac{\dot{a}_6 \cdot (y_F - y_A) + a_6 \cdot \dot{y}_F + \dot{l}_6 \cdot (y_A - y_{G_6})}{l_6}; \dot{z}_{G_6} = \dfrac{\dot{a}_6 \cdot (z_F - z_A) + a_6 \cdot \dot{z}_F + \dot{l}_6 \cdot (z_A - z_{G_6})}{l_6}
\end{cases} \quad (9)$$

$$J_{7SN} = \frac{\frac{1}{2}m_p \cdot R_T^2 + \frac{1}{2}m_p \cdot r_T^2}{2} = \frac{m_p}{4} \cdot (R_T^2 + r_T^2) =$$
$$= \frac{m_p}{4} \cdot \left[R_T^2 + \left(\frac{1}{2}R_T\right)^2\right] = \frac{m_p}{4} \cdot R_T^2 \cdot \left(1 + \frac{1}{4}\right) = \quad (10)$$
$$= \frac{5}{16} \cdot m_p \cdot R_T^2 = \frac{5}{16} \cdot m_p \cdot R^2$$

Where mp shall mean the mass mobile tray 7 (obtained by weighing).

3. THE GEOMETRY AND CINEMATIC OF MOBILE TRAY 7, BY A MATRIX ROTATION METHOD

In Figure 3 is represented mobile plate 7, consisting of an equilateral triangle DEF with the center S. Attach this triangle a system of axs rectangular, mobile, jointly and severally liable with the platform, $x_1Sy_1z_1$.

Known vector $\overline{N}$ coordinates and the coordinates of the pixel S (in relation with the fixed mark considered initially, linked to the fixed platform, be taken as the basis); we know so the co-ordinates of rectangular axis Sz_1, in such a way that can be calculated for a start axis coordinates Sx_1 (relations 11), axis determined by points S, D (known). The co ordinates are obtained vector Sx_1. This, along with the coordinates of the pixel S causes axis Sx_1 (11) (Petrescu, 2009-2013).

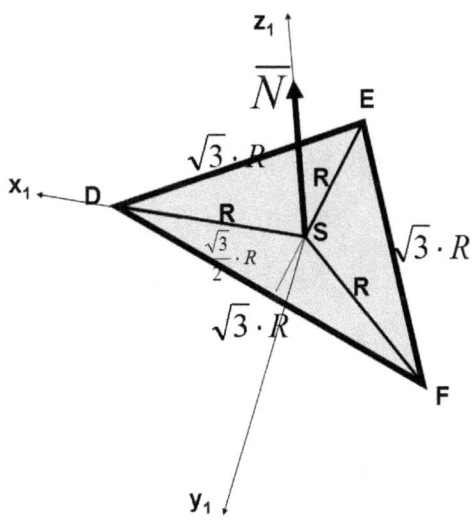

Figure 3. *The geometry and kinematics mobile platform 7*

$$\begin{cases} l_{SD} = \sqrt{(x_D - x_S)^2 + (y_D - y_S)^2 + (z_D - z_S)^2} = \\ = \sqrt{R^2} = R; \quad \alpha_{x_1} = \dfrac{x_D - x_S}{l_{SD}} = \dfrac{x_D - x_S}{R}; \\ \\ \beta_{x_1} = \dfrac{y_D - y_S}{l_{SD}} = \dfrac{y_D - y_S}{R}; \quad \gamma_{x_1} = \dfrac{z_D - z_S}{l_{SD}} = \dfrac{z_D - z_S}{R} \end{cases} \quad (11)$$

By screwing axis $\overrightarrow{Sz_1}$ by (over) axis $\overrightarrow{Sx_1}$, weve axis $\overrightarrow{Sy_1}$ (12). The co-ordinates are thus obtained mobile system $x_1 S y_1 z_1$ (12).

$$\begin{cases} \overrightarrow{Sy_1} = \overrightarrow{Sz_1} \times \overrightarrow{Sx_1} = \begin{vmatrix} \bar{i} & \bar{j} & \bar{k} \\ \alpha & \beta & \gamma \\ \alpha_{x_1} & \beta_{x_1} & \gamma_{x_1} \end{vmatrix} = \\ = (\beta \cdot \gamma_{x_1} - \beta_{x_1} \cdot \gamma) \cdot \bar{i} + (\alpha_{x_1} \cdot \gamma - \alpha \cdot \gamma_{x_1}) \cdot \bar{j} + (\alpha \cdot \beta_{x_1} - \alpha_{x_1} \cdot \beta) \cdot \bar{k} = \\ = \dfrac{\beta \cdot (z_D - z_S) - \gamma \cdot (y_D - y_S)}{R} \cdot \bar{i} + \dfrac{\gamma \cdot (x_D - x_S) - \alpha \cdot (z_D - z_S)}{R} \cdot \bar{j} + \\ + \dfrac{\alpha \cdot (y_D - y_S) - \beta \cdot (x_D - x_S)}{R} \cdot \bar{k} = \alpha_{y_1} \cdot \bar{i} + \beta_{y_1} \cdot \bar{j} + \gamma_{y_1} \cdot \bar{k}; \\ \alpha_{y_1} = \dfrac{\beta \cdot (z_D - z_S) - \gamma \cdot (y_D - y_S)}{R}; \\ \beta_{y_1} = \dfrac{\gamma \cdot (x_D - x_S) - \alpha \cdot (z_D - z_S)}{R}; \Rightarrow [x_1 S y_1 z_1] = \begin{vmatrix} \alpha_{x_1} & \beta_{x_1} & \gamma_{x_1} \\ \alpha_{y_1} & \beta_{y_1} & \gamma_{y_1} \\ \alpha & \beta & \gamma \end{vmatrix} \\ \gamma_{y_1} = \dfrac{\alpha \cdot (y_D - y_S) - \beta \cdot (x_D - x_S)}{R}; \\ \alpha_{x_1} = \dfrac{x_D - x_S}{R}; \quad \alpha_{y_1} = \dfrac{\beta \cdot (z_D - z_S) - \gamma \cdot (y_D - y_S)}{R}; \quad \alpha_{z_1} = \alpha; \\ \beta_{x_1} = \dfrac{y_D - y_S}{R}; \quad \beta_{y_1} = \dfrac{\gamma \cdot (x_D - x_S) - \alpha \cdot (z_D - z_S)}{R}; \quad \beta_{z_1} = \beta; \\ \gamma_{x_1} = \dfrac{z_D - z_S}{R}; \quad \gamma_{y_1} = \dfrac{\alpha \cdot (y_D - y_S) - \beta \cdot (x_D - x_S)}{R}; \quad \gamma_{z_1} = \gamma \end{cases} \quad (12)$$

In Figure 4 is given a positive rotation to axis $\overrightarrow{Sx_1}$ around the axis $\overrightarrow{Sz_1}$ ($\overline{N}$), the angle φ_1.

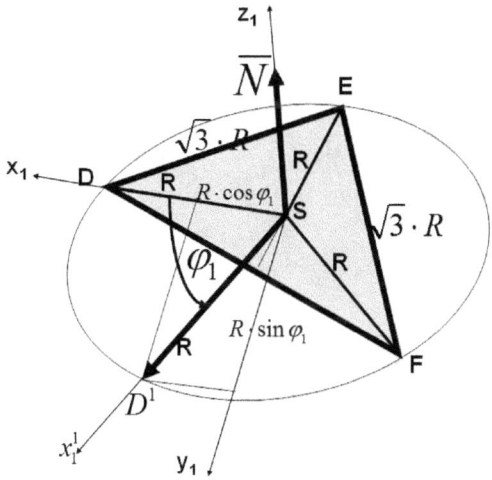

Figure 4. *Rotation around the axis N (within mobile platform)*

Using relations (13) to be written about the system matrix (14), which is determined directly (using rotation matrix) absolute co-ordinates (in accordance with the mark fixed cartesian) of a point D^1 that is part of the plan of mobile top plate. This point moves on the circle of radius R and center S in accordance with rotation imposed by the rotation angle φ_1. Final coordinates are explained in the form (15) (Petrescu, 2009-2013).

$$\begin{cases} \alpha_{x_1} = \dfrac{x_D - x_S}{R}; \; \alpha_{y_1} = \dfrac{\beta \cdot (z_D - z_S) - \gamma \cdot (y_D - y_S)}{R}; \; \alpha_{z_1} = \alpha; \; x_{1D^1} = R \cdot \cos\varphi_1 \\ \beta_{x_1} = \dfrac{y_D - y_S}{R}; \; \beta_{y_1} = \dfrac{\gamma \cdot (x_D - x_S) - \alpha \cdot (z_D - z_S)}{R}; \; \beta_{z_1} = \beta; \; y_{1D^1} = R \cdot \sin\varphi_1 \\ \gamma_{x_1} = \dfrac{z_D - z_S}{R}; \; \gamma_{y_1} = \dfrac{\alpha \cdot (y_D - y_S) - \beta \cdot (x_D - x_S)}{R}; \; \gamma_{z_1} = \gamma; \; z_{1D^1} = 0 \end{cases} \quad (13)$$

$$\begin{aligned}\begin{bmatrix} x_{D^1} \\ y_{D^1} \\ z_{D^1} \end{bmatrix} &= \begin{bmatrix} x_S \\ y_S \\ z_S \end{bmatrix} + \begin{vmatrix} \alpha_{x_1} & \beta_{x_1} & \gamma_{x_1} \\ \alpha_{y_1} & \beta_{y_1} & \gamma_{y_1} \\ \alpha_{z_1} & \beta_{z_1} & \gamma_{z_1} \end{vmatrix} \cdot \begin{bmatrix} x_{1D^1} \\ y_{1D^1} \\ z_{1D^1} \end{bmatrix} = \begin{bmatrix} x_S + \alpha_{x_1} \cdot x_{1D^1} + \beta_{x_1} \cdot y_{1D^1} + \gamma_{x_1} \cdot z_{1D^1} \\ y_S + \alpha_{y_1} \cdot x_{1D^1} + \beta_{y_1} \cdot y_{1D^1} + \gamma_{y_1} \cdot z_{1D^1} \\ z_S + \alpha_{z_1} \cdot x_{1D^1} + \beta_{z_1} \cdot y_{1D^1} + \gamma_{z_1} \cdot z_{1D^1} \end{bmatrix} = \\ &= \begin{bmatrix} x_S + (x_D - x_S) \cdot \cos\varphi_1 + (y_D - y_S) \cdot \sin\varphi_1 \\ y_S + [\beta \cdot (z_D - z_S) - \gamma \cdot (y_D - y_S)] \cdot \cos\varphi_1 + [\gamma \cdot (x_D - x_S) - \alpha \cdot (z_D - z_S)] \cdot \sin\varphi_1 \\ z_S + \alpha \cdot R \cdot \cos\varphi_1 + \beta \cdot R \cdot \sin\varphi_1 \end{bmatrix}\end{aligned} \quad (14)$$

$$\begin{cases} x_{D^1} = x_S + (x_D - x_S)\cdot \cos\varphi_1 + (y_D - y_S)\cdot \sin\varphi_1 \\ \\ y_{D^1} = y_S + [\beta\cdot(z_D - z_S) - \gamma\cdot(y_D - y_S)]\cos\varphi_1 + \\ + [\gamma\cdot(x_D - x_S) - \alpha\cdot(z_D - z_S)]\sin\varphi_1 \\ \\ z_{D^1} = z_S + \alpha\cdot R\cdot \cos\varphi_1 + \beta\cdot R\cdot \sin\varphi_1 \end{cases} \quad (15)$$

Rotation matrix method is used to obtain the point F (for a deduction point coordinates F). Point D shall be superimposed over the point F, if assigns to point D a positive rotation of 120⁰ (relations 16-17). Derive the system (17) and we obtain directly velocities (18) and accelerations (19) of the point F.

$$\begin{cases} x_F = x_{D^1_{120}} = x_S + (x_D - x_S)\cdot \cos 120 + (y_D - y_S)\cdot \sin 120 \\ y_F = y_{D^1_{120}} = y_S + [\beta\cdot(z_D - z_S) - \gamma\cdot(y_D - y_S)]\cdot \cos 120 + \\ + [\gamma\cdot(x_D - x_S) - \alpha\cdot(z_D - z_S)]\cdot \sin 120 \\ z_F = z_{D^1_{120}} = z_S + \alpha\cdot R\cdot \cos 120 + \beta\cdot R\cdot \sin 120 \end{cases} \quad (16)$$

$$\begin{cases} x_F = x_S - \dfrac{1}{2}\cdot(x_D - x_S) + \dfrac{\sqrt{3}}{2}\cdot(y_D - y_S) \\ \\ y_F = y_S - \dfrac{1}{2}\cdot[\beta\cdot(z_D - z_S) - \gamma\cdot(y_D - y_S)] + \\ + \dfrac{\sqrt{3}}{2}\cdot[\gamma\cdot(x_D - x_S) - \alpha\cdot(z_D - z_S)] \\ \\ z_F = z_S - \dfrac{1}{2}\cdot R\cdot \alpha + \dfrac{\sqrt{3}}{2}\cdot R\cdot \beta \end{cases} \quad (17)$$

$$\begin{cases} \dot{x}_F = \dot{x}_S - \dfrac{1}{2}\cdot(\dot{x}_D - \dot{x}_S) + \dfrac{\sqrt{3}}{2}\cdot(\dot{y}_D - \dot{y}_S) \\ \dot{y}_F = \dot{y}_S - \dfrac{1}{2}\cdot[\dot{\beta}\cdot(z_D - z_S) + \beta\cdot(\dot{z}_D - \dot{z}_S) - \dot{\gamma}\cdot(y_D - y_S) - \gamma\cdot(\dot{y}_D - \dot{y}_S)] + \\ \quad + \dfrac{\sqrt{3}}{2}\cdot[\dot{\gamma}\cdot(x_D - x_S) + \gamma\cdot(\dot{x}_D - \dot{x}_S) - \dot{\alpha}\cdot(z_D - z_S) - \alpha\cdot(\dot{z}_D - \dot{z}_S)] \\ \dot{z}_F = \dot{z}_S - \dfrac{1}{2}\cdot R\cdot\dot{\alpha} + \dfrac{\sqrt{3}}{2}\cdot R\cdot\dot{\beta} \end{cases} \quad (18)$$

$$\begin{cases} \ddot{x}_F = \ddot{x}_S - \dfrac{1}{2}\cdot(\ddot{x}_D - \ddot{x}_S) + \dfrac{\sqrt{3}}{2}\cdot(\ddot{y}_D - \ddot{y}_S) \\ \ddot{y}_F = \ddot{y}_S - \dfrac{1}{2}\cdot[\ddot{\beta}\cdot(z_D - z_S) + 2\cdot\dot{\beta}\cdot(\dot{z}_D - \dot{z}_S) + \beta\cdot(\ddot{z}_D - \ddot{z}_S) - \\ \quad - \ddot{\gamma}\cdot(y_D - y_S) - 2\cdot\dot{\gamma}\cdot(\dot{y}_D - \dot{y}_S) - \gamma\cdot(\ddot{y}_D - \ddot{y}_S)] + \\ \quad + \dfrac{\sqrt{3}}{2}\cdot[\ddot{\gamma}\cdot(x_D - x_S) + 2\cdot\dot{\gamma}\cdot(\dot{x}_D - \dot{x}_S) + \gamma\cdot(\ddot{x}_D - \ddot{x}_S) - \\ \quad - \ddot{\alpha}\cdot(z_D - z_S) - 2\cdot\dot{\alpha}\cdot(\dot{z}_D - \dot{z}_S) - \alpha\cdot(\ddot{z}_D - \ddot{z}_S)] \\ \ddot{z}_F = \ddot{z}_S - \dfrac{1}{2}\cdot R\cdot\ddot{\alpha} + \dfrac{\sqrt{3}}{2}\cdot R\cdot\ddot{\beta} \end{cases} \quad (19)$$

For the purpose of determining point coordinates E're still circling the point D with $\varphi_1 = -120^0$ (20). Velocities (21) and accelerations (22) point E is determined by deriving system (20) (Lee, 2013).

$$\begin{cases} x_E = x_S - \dfrac{1}{2}\cdot(x_D - x_S) - \dfrac{\sqrt{3}}{2}\cdot(y_D - y_S) \\ y_E = y_S - \dfrac{1}{2}\cdot[\beta\cdot(z_D - z_S) - \gamma\cdot(y_D - y_S)] - \\ \quad - \dfrac{\sqrt{3}}{2}\cdot[\gamma\cdot(x_D - x_S) - \alpha\cdot(z_D - z_S)] \\ z_E = z_S - \dfrac{1}{2}\cdot R\cdot\alpha - \dfrac{\sqrt{3}}{2}\cdot R\cdot\beta \end{cases} \quad (20)$$

$$\begin{cases} \dot{x}_E = \dot{x}_S - \frac{1}{2}\cdot(\dot{x}_D - \dot{x}_S) - \frac{\sqrt{3}}{2}\cdot(\dot{y}_D - \dot{y}_S) \\ \dot{y}_E = \dot{y}_S - \frac{1}{2}\cdot[\dot{\beta}\cdot(z_D - z_S) + \beta\cdot(\dot{z}_D - \dot{z}_S) - \dot{\gamma}\cdot(y_D - y_S) - \gamma\cdot(\dot{y}_D - \dot{y}_S)] - \\ -\frac{\sqrt{3}}{2}\cdot[\dot{\gamma}\cdot(x_D - x_S) + \gamma\cdot(\dot{x}_D - \dot{x}_S) - \dot{\alpha}\cdot(z_D - z_S) - \alpha\cdot(\dot{z}_D - \dot{z}_S)] \\ \dot{z}_E = \dot{z}_S - \frac{1}{2}\cdot R\cdot\dot{\alpha} - \frac{\sqrt{3}}{2}\cdot R\cdot\dot{\beta} \end{cases} \quad (21)$$

$$\begin{cases} \ddot{x}_E = \ddot{x}_S - \frac{1}{2}\cdot(\ddot{x}_D - \ddot{x}_S) - \frac{\sqrt{3}}{2}\cdot(\ddot{y}_D - \ddot{y}_S) \\ \ddot{y}_E = \ddot{y}_S - \frac{1}{2}\cdot[\ddot{\beta}\cdot(z_D - z_S) + 2\cdot\dot{\beta}\cdot(\dot{z}_D - \dot{z}_S) + \beta\cdot(\ddot{z}_D - \ddot{z}_S) - \\ -\ddot{\gamma}\cdot(y_D - y_S) - 2\cdot\dot{\gamma}\cdot(\dot{y}_D - \dot{y}_S) - \gamma\cdot(\ddot{y}_D - \ddot{y}_S)] - \\ -\frac{\sqrt{3}}{2}\cdot[\ddot{\gamma}\cdot(x_D - x_S) + 2\cdot\dot{\gamma}\cdot(\dot{x}_D - \dot{x}_S) + \gamma\cdot(\ddot{x}_D - \ddot{x}_S) - \\ -\ddot{\alpha}\cdot(z_D - z_S) - 2\cdot\dot{\alpha}\cdot(\dot{z}_D - \dot{z}_S) - \alpha\cdot(\ddot{z}_D - \ddot{z}_S)] \\ \ddot{z}_E = \ddot{z}_S - \frac{1}{2}\cdot R\cdot\ddot{\alpha} - \frac{\sqrt{3}}{2}\cdot R\cdot\ddot{\beta} \end{cases} \quad (22)$$

4. APPLICATIONS

Presented system can be useful in particular to the surgical robot that operate patients who require an accuracy of positioning very high (see figure 5).

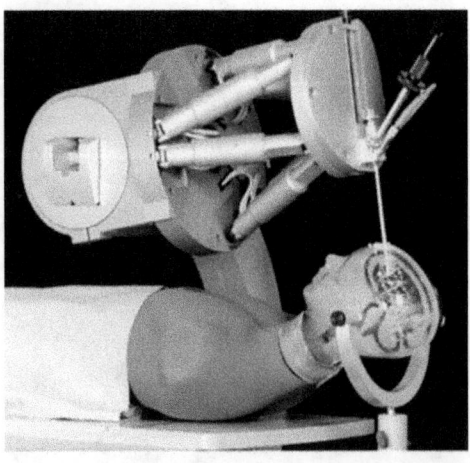

Figure 5. *Surgical robot that operate patients who require an accuracy of positioning very high*

These platforms can position very accurately even very large weights, such as a telescope modern stationary (see Fig. 6).

Figure 6. *A modern stationary telescope positioned by a Stewart system*

Other applications of the platform Stewart are handling and precise positioning of objects large and heavy.

Spatial Stewart platforms may conquer outer space in the future (Flavio de Melo, 2012).

The latest PC-based digital controllers, facilitated by open-interface architecture providing a variety of high-level commands, allow choosing any point in space as the pivot point for the rotation axes by software command (Tang, 2013). Target positions in 6-space are specified in Cartesian coordinates, and the controller transforms them into the required motion-vectors for the individual actuator drives. Any position and any orientation can be entered directly, and the specified target will be reached by a smooth vector motion. The pivot point then remains fixed relative to the platform (Tabaković, 2013).

In addition to the coordinated output of the six hexapod axes, these new hexapod controllers provide two additional axes that can be used to operate rotary stages, linear stages or linear actuators. Some include a macro language for programming and storing command sequences. These controllers feature flexible interfaces, such as TCP/IP interface for remote, network and Internet connection.

New simulation tools are being incorporated for graphical configuration and simulation of hexapods to verify workspace requirements and loads. Such software provides full functionality for creation, modeling, simulation, rendering and playback of hexapod configurations to predict and avoid interference with possible obstacles in the workspace.

With the new design developments that hexapod systems are experiencing, manufacturers and researchers that have a need for extreme high resolutions and high accuracy can now capitalize on them for improvements within their workplace. Hexapod technology has advanced considerably in a few short years, now it is up to industry to embrace these new developments and put them to work to reduce their set-up and processing time, overall production cycle times, and ultimately reduced cost of operation.

5. CONCLUSIONS

The presented method manages to synthesize (in theory) the best option parameters for any desired parallel system. Moving mechanical systems parallel structures are solid, fast, and accurate. Between parallel systems (Wang, 2013) it is to be noticed Stewart platforms, as the oldest systems, fast, solid and precise.

The work outlines a few main elements of Stewart platforms. Begin with the geometry platform, kinematic elements of it, and presented then and a few items of dynamics. Dynamic primary element on it means the determination mechanism kinetic energy of the entire Stewart platforms. It is then in a record tail cinematic mobile by a method dot matrix of rotation.

If a structural mottoelement consists of two moving elements which translates relative, drive train and especially dynamic it is more convenient to represent the mottoelement as a single moving components. We have thus seven moving parts (the six motoelements or feet to which is added mobile platform 7) and one fixed.

Proposed method (in this work) may determine kinematic parameters system position when required the co-ordinates of the endeffector S.

This is clearly a reverse motion (an inverse kinematics) (Lin, 2013).

Method is direct, simple, quick and accurate.

Information display method presented is much simpler and more direct in comparison with methods already known.

REFERENCES

Cao, W., Ding, H., Bin Zi and Ziming Chen (2013). **New Structural Representation and Digital-Analysis Platform for Symmetrical Parallel Mechanisms**, International Journal of Advanced Robotic Systems, Sumeet S Aphale (Ed.), ISBN: 1729-8806, InTech, DOI: 10.5772/56380. Available from:

http://www.intechopen.com/journals/international_journal_of_advanced_robotic_systems/new-structural-representation-and-digital-analysis-platform-for-symmetrical-parallel-mechanisms.

Flavio de Melo, L., Reis Alves, S.F., Rosário, J.M., **Mobile Robot Navigation Modelling, Control and Applications**, in International Review on Modelling and Simulations, April 2012, Vol. 5, N. 2B, pp. 1059-1068. Available from:

http://www.praiseworthyprize.com/IREMOS-latest/IREMOS_vol_5_n_2.html.

Lee, B.J., (2013). **Geometrical Derivation of Differential Kinematics to Calibrate Model Parameters of Flexible Manipulator**, International Journal of Advanced Robotic Systems, Jaime Gallardo-Alvarado, Ramon Rodriguez-Castro (Ed.), ISBN: 1729-8806, InTech, DOI: 10.5772/55592. Available from:

http://www.intechopen.com/journals/international_journal_of_advanced_robotic_systems/geometrical-derivation-of-differential-kinematics-to-calibrate-model-parameters-of-flexible-manipula.

Lin, W., Bing Li, Xiaojun Yang and Dan Zhang (2013). **Modelling and Control of Inverse Dynamics for a 5-DOF Parallel Kinematic Polishing Machine**, International Journal of Advanced Robotic Systems, Sumeet S Aphale (Ed.), ISBN: 1729-8806, InTech, DOI: 10.5772/54966. Available from:

http://www.intechopen.com/journals/international_journal_of_advanced_robotic_systems/modelling-and-control-of-inverse-dynamics-for-a-5-dof-parallel-kinematic-polishing-machine.

Garcia, E.; Jimenez, M.A.; De Santos, P.G.; Armada, M., **The evolution of robotics research**, Robotics & Automation Magazine, IEEE, vol.14, no.1, pp.90,103, March 2007. Available from:

http://ieeexplore.ieee.org/stamp/stamp.jsp?tp=&arnumber=4141037&isnumber=4141014.

He, B., Wang, Z., Li, Q., Xie, H., and Shen, R., (2013). *An Analytic Method for the Kinematics and Dynamics of a Multiple-Backbone Continuum Robot*, IJARS, Patricia Melin (Ed.), ISBN: 1729-8806, InTech, DOI: 10.5772/54051. Available from:

http://www.intechopen.com/journals/international_journal_of_advanced_robotic_systems/an-analytic-method-for-the-kinematics-and-dynamics-of-a-multiple-backbone-continuum-robot.

Liu, H., Zhou, W., Lai, X., and Zhu, S., (2013). *An Efficient Inverse Kinematic Algorithm for a PUMA560-Structured Robot Manipulator*, IJARS, Jaime Gallardo-Alvarado, Ramon Rodrıguez-Castro (Ed.), ISBN: 1729-8806, InTech, DOI: 10.5772/56403. Available from:

http://www.intechopen.com/journals/international_journal_of_advanced_robotic_systems/an-efficient-inverse-kinematic-algorithm-for-a-puma560-structured-robot-manipulator.

Garcia-Murillo, M., Gallardo-Alvarado, J., and Castillo-Castaneda, E., (2013). *Finding the Generalized Forces of a Series-Parallel Manipulator*, IJARS, Jaime Gallardo-Alvarado, Ramon Rodrıguez-Castro (Ed.), ISBN: 1729-8806, InTech, DOI: 10.5772/53824. Available from:

http://www.intechopen.com/journals/international_journal_of_advanced_robotic_systems/finding-the-generalized-forces-of-a-series-parallel-manipulator.

Padula, F., and Perdereau, V., (2013). *An On-Line Path Planner for Industrial Manipulators*, International Journal of Advanced Robotic Systems, Antonio Visioli (Ed.), ISBN: 1729-8806, InTech, DOI: 10.5772/55063. Available from:

http://www.intechopen.com/journals/international_journal_of_advanced_robotic_systems/an-on-line-path-planner-for-industrial-manipulators.

Perumaal, S., and Jawahar, N., (2013). *Automated Trajectory Planner of Industrial Robot for Pick-and-Place Task*, IJARS, Antonio Visioli (Ed.), ISBN: 1729-8806, InTech, DOI: 10.5772/53940. Available from:

http://www.intechopen.com/journals/international_journal_of_advanced_robotic_systems/automated-trajectory-planner-of-industrial-robot-for-pick-and-place-task.

Petrescu, F.I., Petrescu, R.V., *Cinematics of the 3R Dyad*, in journal Engevista, vol. 15, No. 2, (2013), August 2013, ISSN 1415-7314, pp. 118-124. Available from:

http://www.uff.br/engevista/seer/index.php/engevista/article/view/376.

Petrescu, F.I., Petrescu, R.V., *Kinematics of the Planar Quadrilateral Mechanism*, in journal Engevista, vol. 14, No. 3, (2012), December 2012, ISSN 1415-7314, pp. 345-348. Available from:

http://www.uff.br/engevista/seer/index.php/engevista/article/view/377.

Petrescu, F.I., Petrescu, R.V., *Mecatronica – Sisteme Seriale si Paralele*, Create Space publisher, USA, March 2012, ISBN 978-1-4750-6613-5, 128 pages, Romanian edition.

Petrescu, F.I, Petrescu, R.V, *Mechanical Systems, Serial and Parallel – Course* (in romanian), LULU Publisher, London, UK, February 2011, 124 pages, ISBN 978-1-4466-0039-9, Romanian edition.

Petrescu, F.I., Grecu, B., Comanescu, A., Petrescu, R.V., *Some Mechanical Design Elements*. In the 3rd International Conference on Computational Mechanics and Virtual Engineering, COMEC 2009, Brașov, October 2009, ISBN 978-973-598-572-1, Edit. UTB, pp. 520-525.

Reddy, P., Shihabudheen K.V., Jacob, J., *Precise Non Linear Modeling of Flexible Link Flexible Joint Manipulator*, in International Review on Modelling and Simulations, June 2012, Vol. 5, N. 3B, pp. 1368-1374. Available from:

http://www.praiseworthyprize.com/IREMOS-latest/IREMOS_vol_5_n_3.html.

Tabakovic, S., Milan Zeljkovic, Ratko Gatalo and Aleksandar Zivkovic (2013). *Program Suite for Conceptual Designing of Parallel Mechanism-Based Robots and Machine Tools*, International Journal of Advanced Robotic Systems, Sumeet S Aphale (Ed.), ISBN: 1729-8806, InTech, DOI: 10.5772/56633. Available from:

http://www.intechopen.com/journals/international_journal_of_advanced_robotic_systems/program-suite-for-conceptual-designing-of-parallel-mechanism-based-robots-and-machine-tools.

Tang, X., Sun, D., and Shao, Z., (2013). *The Structure and Dimensional Design of a Reconfigurable PKM*, IJARS, Sumeet S Aphale (Ed.), ISBN: 1729-8806, InTech, DOI: 10.5772/54696. Available from:

http://www.intechopen.com/journals/international_journal_of_advanced_robotic_systems/the-structure-and-dimensional-design-of-a-reconfigurable-pkm.

Tong, G., Gu, J., and Xie, W., (2013). **Virtual Entity-Based Rapid Prototype for Design and Simulation of Humanoid Robots**, International Journal of Advanced Robotic Systems, Guangming Xie (Ed.), ISBN: 1729-8806, InTech, DOI: 10.5772/55936. Available from:

http://www.intechopen.com/journals/international_journal_of_advanced_robotic_systems/virtual-entity-based-rapid-prototype-for-design-and-simulation-of-humanoid-robots.

Wang, K., Luo, M., Mei, T., Zhao, J., and Cao, Y., (2013). **Dynamics Analysis of a Three-DOF Planar Serial-Parallel Mechanism for Active Dynamic Balancing with Respect to a Given Trajectory**, International Journal of Advanced Robotic Systems, Sumeet S Aphale (Ed.), ISBN: 1729-8806, InTech, DOI: 10.5772/54201. Available from:

http://www.intechopen.com/journals/international_journal_of_advanced_robotic_systems/dynamics-analysis-of-a-three-dof-planar-serial-parallel-mechanism-for-active-dynamic-balancing-with-.